Pests in the Urban Biome

_____

Pests in the Urban Biome

William H. Robinson

CABI

CABI is a trading name of CAB International

CABI	CABI
Nosworthy Way	200 Portland Street
Wallingford	Boston
Oxfordshire OX10 8DE	MA 02114
UK	USA
Tel: +44 (0)1491 832111	T: +1 (617)682-9015
E-mail: info@cabi.org	E-mail: cabi-nao@cabi.org
Website: www.cabi.org	

The views expressed in this publication are those of the author(s) and do not necessarily represent those of, and should not be attributed to, CAB International (CABI). Any images, figures and tables not otherwise attributed are the author(s)' own. References to internet websites (URLs) were accurate at the time of writing.

CAB International and, where different, the copyright owner shall not be liable for technical or other errors or omissions contained herein. The information is supplied without obligation and on the understanding that any person who acts upon it, or otherwise changes their position in reliance thereon, does so entirely at their own risk. Information supplied is neither intended nor implied to be a substitute for professional advice. The reader/user accepts all risks and responsibility for losses, damages, costs and other consequences resulting directly or indirectly from using this information.

CABI's Terms and Conditions, including its full disclaimer, may be found at https://www.cabidigital-library.org/terms-and-conditions.

A catalogue record for this book is available from the British Library, London, UK.

ISBN-13: 9781800626393 (hardback)
      9781800626409 (ePDF)
      9781800626416 (ePub)

DOI: 10.1079/9781800626416.0000

Commissioning Editor: Ward Cooper
Editorial Assistant: Theresa Regueira
Production Editor: Rosie Hayden

Typeset by Exeter Premedia Services Pvt Ltd, Chennai, India
Printed in the USA

Contents

Preface: The Urban Biologist

———————————

Urban biologists are the new generation of scientists engaged in research and education on pest species in the urban biome. They will establish a discipline that includes pest species and the environmental conditions and physical features of this global biome. Research will be on the habitats, harborages, and resources that determine the biological and behavioral fitness of urban species. These links are critical to survival in the urban biome but are vulnerable to gradual climatic change or a sudden invasive challenge. Synanthropic status is unique but not a long-term guarantee, as the food resource or other habitat conditions may be changed. The dynamics of fitness to the urban biome will be a key part of the next paradigm of research on urban pests.

This book takes a new view of pest species in the urban biome. It is written as an introduction for a broad audience of biologists and non-biologists. There is extensive use of pictures and graphics to explain pest biology, and limited use of technical terms. The objective is to support a discipline that integrates species with habitats and encourages interdisciplinary research. An expansion of topics and species research will draw a circle that includes other disciplines and subjects, and the name "urban biologist" would recognize this inclusion. The next generation of scientists and their research on the urban biome will be multi-disciplined. The number and scope of multi-disciplined studies must begin to match the importance of this subject.

The early chapters provide a basic introduction to the biome concept, followed by a review of landscape and environmental features. The topics covered in these chapters are linked directly or indirectly to the development and success of urban species. The pest species are divided in chapters that emphasize their habitat fitness, the importance of their harborage, their reservoir populations, the potential ecological traps for synanthropic species, and the influence of a changing climate on the distribution of important vector species.

William H. Robinson
Christiansburg, Virginia
August 2025

Acknowledgments

————————————

I am thankful to all those who gave encouragement and provided content, corrections, and images that improved this book. Rudy Plarre, BAM Federal Institute for Materials Research and Testing, Berlin, Germany, read it carefully and kept me on-message, and, with his insights, strengthened the synanthropic trap concept. Mariah Valente Baggio, Technical Director, Pestmaster, Reno, NV, USA, provided insightful perspectives on key chapters. Shu-Ping Tseng, Department of Entomology, National Taiwan University, Taipei City, Taiwan, considered the book's concept, audience, and content as a textbook for a wide audience. Urban entomology students, including Tzong-Han Lin and Chungswat, provided images of cockroach legs. Brian Forschler, Department of Entomology, University of Georgia, Athens, Georgia, USA, tracked the terminology to keep it current. Maria Nemeth read the chapters as a non-biologist to keep it interesting and concise. The commitment and enthusiasm of the urban biologists participating in the 30 years of meetings of the International Conference on Urban Pests (ICUP) provided content and stimulus for this book.

Chapter images

There are 181 images in this book. They were selected, designed, or prepared specifically to contribute to the text they accompany on the page. About 132 of the images were considered public domain and resized or edited to fit a topic. About 50 images, including the insect life cycles, harborage graphics, and photographs, were taken or created by the author.

1 Anthropogenic Biomes

In 1939, Frederick Clements and Victor Shelford proposed that plants and animals form inter-related communities that acted as organic units or biomes. This concept was expanded to include environmental factors, such as temperature and precipitation. The traditional biome definition refers to natural habitats without influence from humans. Now there are few natural biomes that have not had human contact or change. In a 2008 publication, Erle Ellis and Navin Ramankutty introduced the concept of anthropogenic biomes or anthromes (human biomes). They considered the global anthropogenic biomes composed of human systems, with natural ecosystems embedded within them.

Urban Biome

Biomes around the world struggle with the excesses of size and spread of the human population. Urbanization influences conditions at the local level and climate change influences conditions at the global level. The urban biome is the dominant and most influential biome on the planet. This biogeographical space is occupied by 65% of the world's population, and they live at a density of 186 individuals per square kilometer. The landscape of this biome begins at the city center (Fig. 1.1), which has a cover of buildings and roads, and extends to the geographic edge with a mix of buildings and green and blue space. The size and geographic location of cities can determine the urban heat island they create, and the amount of green space and blue space distinguishes some old from new cities. The sprawl of large cities into the surrounding area and toward adjacent cities has created the land form of megalopolis around the world.

As the urban biome developed, it gradually changed the original natural or agricultural topography it was a part of. Original forested areas and large lakes, blue spaces, and stands of trees or shrubs became green spaces or parks in urban neighborhoods. Some habitats and the plants and animals they supported were significantly reduced or eliminated. Large green and blue spaces were repurposed, and new habitats, such as Central Park in New York City, were created.

Extensive sewer and stormwater drainage systems in cities became an underground habitat connecting all regions of the city and suburbs. This unique habitat supports numerous species adapted to the conditions. Whether new or relic habitats, above ground on sidewalks or ledges of buildings, or below ground in drainage systems or transportation tunnels, the urban biome provides conditions and food resources for vertebrate and invertebrate species to become established.

Local conditions and resources determine the distribution and abundance of species in the urban biome. Some species are seasonal and linked to a specific habitat or food source, such as mosquitoes or ticks in blue and green spaces. Others are linked to conditions and food resources, such as house flies breeding in displaced food waste and garbage. The insects attracted to street lights and other lights at night can create localized problems but these may be resolved with the increased use of LED (light-emitting diode) bulbs. The synanthropic species

DOI: 10.1079/9781800626416.0001

Fig. 1.1. Urban biome landscape. Credit: Pixabay/CC0 Public Domain

seem broadly or specifically adapted to urban conditions and infrastructure. These species have global distribution, such as American cockroaches in sewers, pharaoh ants in houses, and brown rats and pigeons which are adapted to the structure of buildings and the abundance of human food waste.

Indoor biome

The indoor biome can be considered a subset of the urban biome and composed of the living and working space in residential and commercial buildings. It is estimated to be 6% of the global land area. The concept of the indoor biome is based on it being a defined space with defined conditions that support a community of organisms in addition to humans. This is a built space where humans spend 90% of their lives living and working. That level of habitation and activity, which often includes pets, influences the conditions. The building materials and construction quality, the physical location, and the surrounding climate also have a part. The interior and external design determines the connection between the indoor and outdoor

environment on a daily and seasonal basis. The diversity of species that are expected to occur indoors are linked to the surrounding landscape, and the routes available to enter indoors.

The key features of the indoor biome that determine the residency of invading organisms, especially insects and other arthropods, are moisture (relative humidity), temperature stability, human activity, and building materials. Protection from water loss has survival value for insects indoors, and the successful species are adapted or have adjusted to the temperature and humidity. Human activity often provides pests suitable food and immediate access to indoors. Building materials and construction features can provide gaps and spaces to become shelter and harborages for arthropods.

Isobenefit Urbanism

This concept proposes cities where key points, such as natural areas, job locations, shops, and recreational, medical, and cultural institutions, can be reached by walking or cycling (Fig. 1.2).

Fig. 1.2. Isobenefit urban scene. Credit: Pixabay/CC0 Public Domain

The objective is to have these beneficial sites within a mile (1.6 km) of city residents. It builds on the concept of "walking cities" in which the nearest services, workplaces, and green or blue spaces are within a 1-mile walking distance. This new form would be implemented by building new cities, or by designing the expansion of existing ones to fit the isobenefit objective.

Urban blue and green spaces provide natural areas within cities that contribute to the better health and wellbeing of urban residents. However, if degraded, these ecosystems pose a potential risk for human health. Understanding the role of natural areas in urban resident health and wellbeing is crucial for developing isobenefit cities, particularly for older adults.

2 Landscape and Structure

The term "urban" is associated with the city, but when used in the context of the urban environment it extends to plant and animal communities in cities and suburbs. There is a continuum of inhabited sites and human activity from the rural farmhouse to the metropolitan office building, and the division between urban, suburban, and rural areas is often indistinct. Within the urban environment there are distinct areas of modification and changes in the physical landscape and biotic communities.

Urban Modifications

The urban landscape has evolved and the changes are recognized as distinct land forms or populations densities. An interface with agriculture or natural areas can be distinct because of land use. A large megalopolis may include several urban core zones. Suburbs or a low-income fringe can define the edge of a megalopolis.

Suburbia

Development of what is known as "suburbia" began in the 1800s with people moving to the perimeter of the industrial cities. By the 1960s, major cities in the industrial countries had a distinct suburban perimeter, and a distinct commercial core. Suburbia is generally considered to be composed of planned communities and structured green space. The creation of the 21st-century megalopolis was aided by the phenomenon of "counter-urbanization" in which large numbers of urban residents moved to the suburbs. This is the reverse of urbanization in which the city attracted people, and it has contributed to metropolis growth. This ex-urban sprawl or extension of cities started in the 1970s and continues in developed countries.

Megalopolis

Urbanization continues around the world through urban sprawl and the creation of the megacity or megalopolis. This is a gradual process in which suburban land use spreads into peripheral farmland and natural areas. The first of these megacities formed in the North-east corridor along the Atlantic coast of the USA (Fig. 2.1). It includes five core cities: Boston, New York, Philadelphia, Baltimore, and Washington, DC, and regions in 12 states. This is a global phenomenon, with megacities located in Latin America, Europe, Africa, and Asia. There are 26 megacities with populations exceeding 20 million around the world, and 30 proto-megacities, with populations over 15 million and growing rapidly.

The term "Blue Banana" is used to describe an economic zone and continuous urban area in western and central Europe. The demarcated territory has a shape that suggests a banana (Fig. 2.2). It extends from north-west England to northern Italy. It is inhabited by about 100 million people with a population density of about 400 people per square km. The boundaries of the Blue Banana include areas with high levels of urbanization and population density but also features of temperate climate, fertile soils, and mineral resources. Economic centers and intense industrial centers grouped in this geographic band distinguish it from other regions of the continent.

© William H. Robinson 2026. *Pests in the Urban Biome* (W.H. Robinson)
DOI: 10.1079/9781800626416.0002

Fig. 2.1. North-eastern US megalopolis. Credit: US Census Bureau.

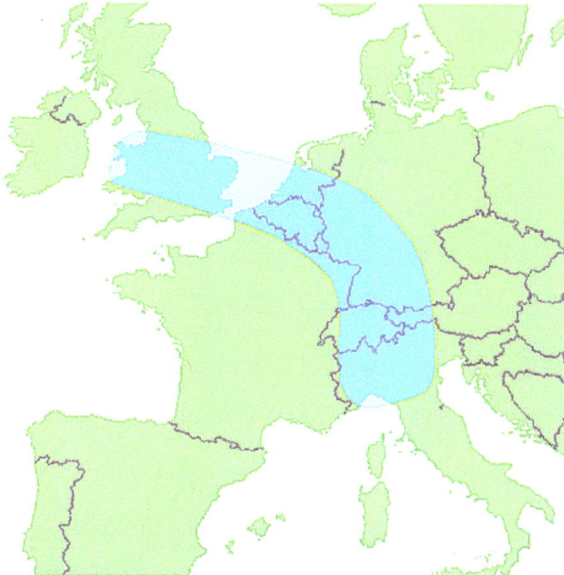

Fig. 2.2. European economic and urban zone, the blue banana. Credit: Wikipedia/CC0 Public Domain.

Fig. 2.3. Sprawl around city perimeter. Credit: Pixabay/CC0 Public Domain.

Sprawl is a universal feature of cities and suburban growth. It is a pattern of population expansion into surrounding open spaces. It may be a planned growth or an unplanned spread. In the USA, the encroachment of forested or agricultural lands results in a loss of more than 2000 green hectares each day in the 21st century, on average. This expansion is often unplanned, and the result is fragmentation of the landscape (Fig. 2.3). Native and introduced species can utilize a fragmented landscape and achieve large populations, while populations of native species are reduced. When urban peripheral land is "consumed" faster than the urban population growth, the metropolitan area is defined as sprawling.

Donut effect

Metropolitan areas around the world have experienced the population density phenomenon of the "donut effect." This refers to the migration of residential units and small businesses from city central districts to the surrounding areas. Since the 2020 COVID-19 pandemic, 12 of the largest cities in the USA have cumulatively lost 8% of their downtown dwellers, and 60% of those households moved to nearby suburbs.

Low-income fringe

At the periphery of cities in some developing countries are zones of dense, unplanned, and often impoverished housing. These sites vary from country to country, but they are an established feature of major cities and represent 20–30% of the new urban housing in the world. Most new housing in developing countries is built on unclaimed land and involves substandard housing with limited clean water and waste removal. These are habitats for invertebrate disease vectors with a flight range to a large portion of the city's population.

Urban core

This is the most developed part of the urban built-up zone. The core contains about 60% of the city surface area and consists of hard surfaces, such as sidewalks and parking sites, and the exterior surfaces of built structures. This is the human-built landscape of the city and is characterized by an uneven distribution of exposed soil, which is around street trees and vacant lots, and limited green space. It is dominated by the hard surfaces that are exposed to sunlight and hold that heat

Fig. 2.4. Interface of agriculture and urban zones. Credit: Pixabay/CC0 Public Domain.

energy when the sun sets. The existing plants are selected and maintained by human activity, and the insects, rodents, and birds use habitats and food that is different from their natural preference.

Edge cities

These residential and commercial sites are at the perimeter of large metropolitan areas. They are structured suburbs that are not dependent on the central city but are self-sufficient cities. Edge cities are marked by industrial parks, high-rise office buildings, retail complexes, and apartment and condominium districts. They are in all regions of the USA, and some metro-areas around the world, and are becoming typical elements of the urban landscape. They have specific economic, socio-cultural, and spatial conditions that are supported by highway and rail connections.

Agriculture interface

An urban interface with agriculture often occurs when suburban sprawl brings new residential and commercial land close to farm land or animal feed lots (Fig. 2.4). Dairy cattle, swine, beef livestock, and poultry operations are often encroached upon as the suburban ring of cities spreads. Manure produced at these operations can support large populations of house flies and stable flies, and they can disperse several miles into non-farm locations.

Natural area interface

The urban interface with undisturbed or natural areas occurs when residential housing developments are built close to land set aside or preserved as a natural site. Undisturbed areas may provide reservoir populations for household and peridomestic pest species, including yellowjackets and carpenter bees, carpenter ants, subterranean termites, and species of ticks. Animal populations increase the potential of spreading arthropod-borne diseases to the residents using the natural areas as convenient recreation sites for people and pets.

Geo-Ecological Habitats

The structural complexity of cities includes features that provide harborage and food for arthropods and other animals. Parks, recreation areas, and other

Fig. 2.5. Landscape complexity of urban environment. Credit: Pixabay/CC0 Public Domain.

green spaces have natural habitats for vertebrates and invertebrates. The system of stormwater and sewer pipes provides artificial habitats for insects and small animals. Garbage collection points and landfills are features of urban environments, and they provide habitats and human food waste for arthropods, rodents, and pest birds.

Built structures

Enclosed spaces in multi-story buildings for living and working are concentrated in the urban core (Fig. 2.5). These spaces are a controlled-temperature and humid environment with little fluctuation over 12 months. Activities and materials are separated into cooking, sleeping, toilet, and storage spaces. The global homogenization of the features and format of the living and working space results in structural similarities across cultures and continents. Large parking garages, sports stadia, and outdoor performance and event areas are a feature of large and small

urban areas. Although the use is usually temporary and seasonal, the activities are accompanied by food preparation and accumulations of human food waste. When unattended, these sites can provide food and harborage for pigeons and other pest birds, and rodents.

Green space

Many cities have been designed to include space for parks, peripheral green belts, or slightly forested sites. These areas break the monotony of residential and commercial buildings, influence local temperature and humidity, and provide recreation sites and contact with nature (Fig. 2.6). Urban green space fulfills environmental roles, including reducing noise pollution and absorbing atmospheric gases and particulate pollutants. Linear green spaces include green corridors, such as rights-of-way, riverside walks and bicycle routes, and parks between buildings.

Parks and vegetation corridors, small, forested sites, and grassy areas reduce surrounding

Fig. 2.6. Linear urban green space. Credit: Pixabay/CC0 Public Domain.

urban temperatures. Parks moderate high temperatures from the urban heat island effect by providing shade, they enhance wind patterns, and the cooler air over parks replaces warmer air in adjacent neighborhoods. Urban parks encourage visiting and regular activity of nearby residents. They also encourage activity with companion dogs and cats. However, the dense vegetation and humid conditions are suitable for arthropod disease vectors, such as mosquitoes and ticks. The vertebrate hosts for these vectors often establish populations in parks, including rabbits, deer, chipmunks, squirrels, mice, and voles.

The "3-30-300" rule is a sustainable measure that encourages every house, school, and workplace to have a view of at least three trees, be in a neighborhood with at least 30% canopy cover, and be within 300 m of a park. Depression, anxiety, obesity, and heatstroke are reported to be more prevalent in urban areas that lack this level of access to a shady tree canopy and green open spaces. Older adults experience the feeling of wellbeing and stress reduction while spending time in green spaces. Some research finds that the health benefits of time in natural environments are dose-dependent and suggests that repeated exposure brings the best benefit. Research suggests that there is not one green space type or characteristic that works best for everyone. There may be a need for variety in green space types to suit different users. It is not only urban green spaces that appeared to matter, but also street greenery, and trees (Fig. 2.7).

Pocket parks

Small parks generally characterize the green space of large and small cities. These spaces can provide a cooling effect that correlates with their size. For example, small green spaces with an area of 300 m^2 can result in 1°C temperature reduction, slightly larger parks with an area of 650 m^2 can reduce the temperature by up to 2°C, and a 1500 m^2 green space can reduce the temperature by up to 3.6°C. The shape of small plots of green space influences cooling; for example polygonal-shaped parks with varied vegetation cover can reduce the temperature by up to 4°C.

Blue space

Blue spaces are any urban surface water, such as ponds, lakes, streams and rivers, and in some cases the ocean (Fig. 2.8). Blue spaces offer temperature

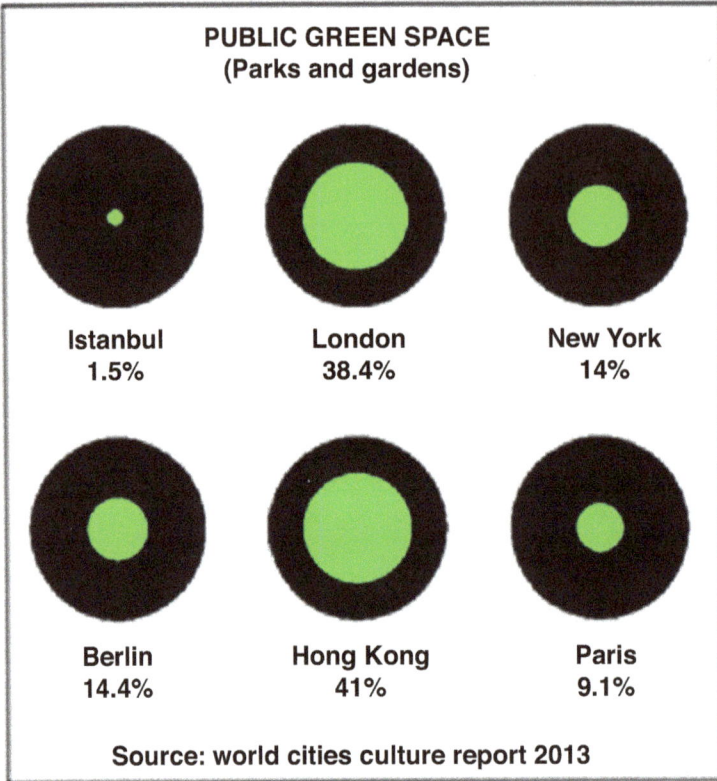

PUBLIC GREEN SPACE
(Parks and gardens)

Istanbul
1.5%

London
38.4%

New York
14%

Berlin
14.4%

Hong Kong
41%

Paris
9.1%

Source: world cities culture report 2013

Fig. 2.7. Green space in major cities. Credit: World Cities Culture Report 2013.

Fig. 2.8. Blue space in urban landscape. Credit: Pixabay/CC0 Public Domain.

regulation processes, absorbing heat during the day when air temperatures exceed water temperatures and releasing heat during the night when water temperatures exceed air temperatures. Urban blue space sites can have an average cooling effect of 2.5°C during May to October.

Urban blue spaces contribute to reducing urban heat stress. The benefits target daytime urban temperatures and provide a cooling effect. Cooling is influenced by the size of these spaces and the distance between them. Large water bodies provide the greatest cooling effect in zones adjacent to their boundaries and in downwind areas. The size and length of the downwind spread is dependent on the wind velocity, with relatively cooler air originating from a large urban water body transported by winds to generate plumes of cool air. Several small water bodies distributed within an urban area generate lower temperature effects than a single water body of similar total volume. The heat-reducing properties of rivers and streams are influenced by fluid flow. Their flow enables them to carry absorbed radiation downstream and release energy externally to the urban system.

The capacity of urban blue space to mitigate urban heat islands has also been shown to relate to positive human health impacts such as reductions in heat-related mortality amongst people who live within 4 km of water. Introducing new and maintaining established water bodies within a city is difficult, but the combination of urban blue and green provides the highest benefit. New green areas with planned water bodies will make an important contribution to the urban climatic environment and the health of urban residents.

Brown space

Brown spaces, also known as brownfields, are any previously developed or developed land that is not currently in use (Fig. 2.9). The term "brown space" is also applied to land previously used for industrial or commercial purposes, and may be linked to pollution or soil contamination due to hazardous waste. These sites waste valuable urban space because they could be providing some benefit, such as a vegetation

Fig. 2.9. Undeveloped urban brown space. Credit: Pixabay/CC0 Public Domain.

cover that contributes to water retention and reduces runoff. Reclamation of brownfields has an aesthetic benefit to the community and improves property values for the surrounding neighborhoods. Naturalization of these sites presents an opportunity to create new green space in urban areas, and allows native plants, insects, birds, and other wildlife to become reestablished.

Unfinished and abandoned space

Unfinished construction projects are damaging to urban space; starting with the land preparation and construction, then again when they remain unfinished and abandoned. These buildings can become vertical habitats for plants and animals (Fig. 2.10). They become public health hot spots when they support rodent, bird, and insect vector populations. The cost of removing and reclaiming them is often economically prohibitive. But the cost of their ongoing presence to neighborhoods, communities, and the environment is considerable.

Underground sewers and subways

These urban underground systems include the network of infrastructure built beneath the surface, including subway train systems ("metros") (Fig. 2.11), water pipes and sewage ducts, electrical cables, and pedestrian walkways beneath roads. Sewer systems are a network of pipes that transport wastewater from buildings through sanitary sewer pipes to a wastewater treatment facility. Stormwater drainage pipes carry away water following rain or snow events. Water from streets and sidewalks flows into the underground pipes through grated openings along the curb and at street corners. Because of their position along streets, they often become clogged with trash and food waste and become a breeding site for insects and rodents.

Metros or subways are the underground railway systems used to transport large numbers of passengers within urban and suburban areas. Subways are large tunnels that accommodate several train lines in one space. Underground walkways serve pedestrians in dense urban core areas. These tunnels often have shops selling food and drinks and that remain open for long hours. Water from these facilities and the discarded food containers provide habitats for mosquitoes, cockroaches, and brown rats. The conditions allow these pests to remain active year-round.

Solid waste disposal and landfills

Waste produced by residential and commercial sources is transferred to a landfill, a site designed for and dedicated to waste disposal (Fig. 2.12). Municipal solid waste originates from daily activity in commercial locations, and it contains 10–50% wet and putrescible organic material.

Fig. 2.10. Abandoned urban housing complex. Credit: Pixabay/CC0 Public Domain.

Fig. 2.11. Underground urban subway or metro. Credit: Pixabay/CC0 Public Domain.

Fig. 2.12. Active urban landfill. Credit: Pixabay/CC0 Public Domain.

This high organic content is suitable food and harborage for insects, pest birds, and rodents. Populations of gulls in cities are closely tied to availability of human food waste in landfills. House flies and bottle flies breed in landfills. As many as 2 million house flies can be in a municipal landfill as eggs, larvae, or pupae during a peak month in summer.

Additional Reading

Anderson, M., Ware, R. and Murphy, R.G. (2005) Population survey of mosquito species in proximity to an international airport and other urban sites in the UK. In: Lee, C.-Y. and Robinson, W.H. (eds) *Proceedings of the Fifth International Conference on Urban Pests*, ICUP, Singapore, pp. 271–274. Available at: https://icup.org.uk/media/wimj3lzd/icup042.pdf (accessed 29 September 2025).

Boase, C. (1999) Trends in urban refuse disposal: A pest's perspective. In: Robinson, W.H., Rettich, F. and Rambo, G.W. (eds) *Proceedings of the Third International Conference on Urban Pests*, ICUP, Prague,

Czech Republic, pp. 83–98. Available at: https://icup.org.uk/media/ftxhtjfb/icup400.pdf (accessed 29 September 2025).

Boase, C. (2017) House flies: Regulations and unintended consequences. In: Davies, M.P., Pfeiffer, C. and Robinson, W.H. (eds) *Proceedings of the Ninth International Conference on Urban Pests*, ICUP, Birmingham, UK, pp. 53–59. Available at: https://icup.org.uk/media/434fou5y/icup1183.pdf (accessed 29 September 2025).

Caimi, M., Cassani, S., Girgenti, P. and Süss, L. (2005) Evaluation of arthropod fauna in an urban waste treatment plant. In: Lee, C.-Y. and Robinson, W.H. (eds) *Proceedings of the Fifth International Conference on Urban Pests*, ICUP, Singapore, p. 499. Available at: https://icup.org.uk/media/0i2aye h4/icup086.pdf (accessed 29 September 2025).

Hirabayashi, K., Tanizaki, S. and Yamamoto, M. (2008) Chironomid (Diptera, Chironomidae) fauna in a filtration plant in Japan. In: Robinson, W.H. and Bajomi, D. (eds) *Proceedings of the Sixth International Conference on Urban Pests*, ICUP, Budapest, Hungary, pp. 187–192. Available at: https://icup.org.u k/media/vqqfhb2i/icup901.pdf (accessed 29 September 2025).

Reierson, D.A., Rust, M.K. and Paine, E. (2005) Control of American Cockroaches (Dictyoptera: Blattidae) in sewer systems. In: Lee, C.-Y. and Robinson, W.H. (eds) *Proceedings of the Fifth International Conference on Urban Pests*, ICUP, Singapore, pp. 141–148. Available at: https://icup.org.uk/media/ tbgpmuyo/icup023.pdf (accessed 29 September 2025).

Stejskal, V. (2002) Metapopulation concept and the persistence of urban pests in buildings. In: Jones, S., Zhai, J. and Robinson, W. (eds) *Proceedings of the Fourth International Conference on Urban Pests*, ICUP, Charleston, USA, pp. 75–85. Available at: https://icup.org.uk/media/obydtsvx/icup207 .pdf (accessed 29 September 2025).

Sûss, L., Cassani, S., Serra, B. and Caimi, M. (1999) Integrated pest management for control of the house fly *Musca domestica* (L.) (Diptera: Muscidae) in an urban solid waste treatment plant. In: Robinson, W.H., Rettich, F. and Rambo, G.W. (eds) *Proceedings of the Third International Conference on Urban Pests*, ICUP, Prague, Czech Republic, pp. 261–267. Available at: https://icup.org.uk/media/pxrpiz15 /icup426.pdf (accessed 29 September 2025).

3 Environmental Features

Urbanization affects the components of the environment. The concentrations of heat-absorbing surfaces of roads and parking lots, the limited green space and open soil, and particulate matter in the air result in cities having a climate different from the surrounding landscape. Climatic changes bring seasonal temperature highs and lows as well as changes in the intensity and direction of wind around buildings and in the amount of rainfall and runoff.

Hard Surfaces

Up to 33% of the land surface in cities is in the form of roads, sidewalks, and parking lots. A nearly equal proportion is taken up by buildings and other built structures, with the result that 60–70% of urban areas in modern cities consists of surfaces formed from nonporous materials. Only the remaining third of urban surface can be considered porous for water circulation and water vapor exchange.

Hard surfaces accept more heat energy in less time than an equal area of soil. By the end of the day, brick or concrete surfaces will have stored more heat than an equal surface of soil. Hard surfaces release heat about three times as fast as it is released by moist soil. Urban buildings have a windbreak effect on the prevailing wind, and this may reduce the amount of heat that is carried away. Street trees can lower surrounding air temperatures by up to 5.6°C and road surface temperatures by up to 27.5°C.

Major roads and highways

Metropolitan areas are generally centered around major road systems, and multiple-lane highways are a standard feature (Fig. 3.1). As

Fig. 3.1. Urban highway complex. Credit: Pixabay/CC0 Public Domain.

© William H. Robinson 2026. *Pests in the Urban Biome* (W.H. Robinson)
DOI: 10.1079/9781800626416.0003

URBAN HEAT ISLAND PROFILE

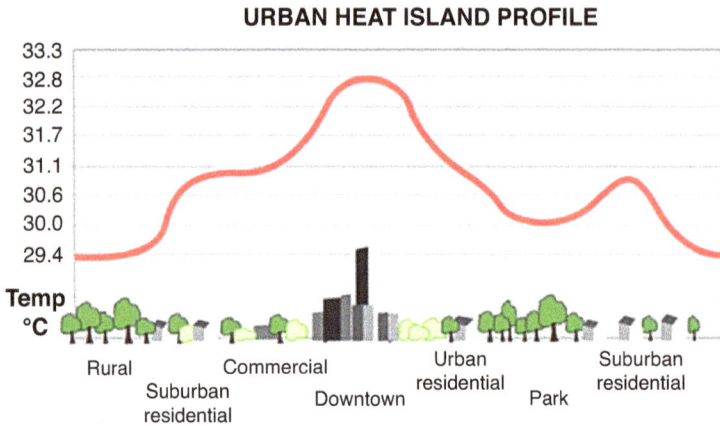

Fig. 3.2. Urban heat island (UHI). Credit: Wikipedia/CC0 Public Domain.

these are initially built and expanded, they may encroach on established animal territories or seasonal movement routes. However, roads and highways do not completely restrict the foraging territory or dispersal routes of animals in urban habitats. Some terrestrial species adjust to these barriers over time and learn ways to travel around them. Animal populations that have high exposure to highways and have been in place for many years show signs of adaptation to the road noise and speeding vehicles. For example, bats roosting near busy highways adjust to the traffic sounds. Culverts under small and major road systems are commonly used for dispersal by rodents and other animals. White-footed mice, chipmunks, and grey squirrels have demonstrated a unique ability to acclimate to these channels, and to maintain successful populations on both sides of a highway.

Urban Islands

The urban island concept links specific environmental conditions, such as temperature and humidity, air pollution, and artificial light, to a defined urban area. This helps to separate the conditions from other features and show their importance to the whole urban ecosystem.

Urban heat island

Buildings and other features add to the three-dimensional complexity of cities. The result is a rise in the mean temperature, forming what is called an urban heat island (UHI) (Fig. 3.2). This "island" results from the reduced amount of evaporative cooling, heat retained by surfaces, and heat produced by vehicles and machines. There is a limited range of daily high and low temperatures in the UHI. Summer nights in suburban areas may be cool, but city temperatures may be only a few degrees lower at midnight than at sundown. The temperature difference between an urban area and the rural surrounding landscape can reach 12°C.

Heat islands can occur year-round during day or night. The maximum heat island effect occurs approximately 3–5 hour after sunset. Seasonal variations in weather patterns can also affect heat island frequency and magnitude. Heat islands of cities located in the mid latitudes usually are strongest in the summer or winter seasons. Those that occur during summer are more concerning because of elevated air pollution, and heat stress-related mortality and illness. Climate change will likely lead to more frequent and longer heat waves. Large cities could experience more days with temperatures above 30°C with increased greenhouse emissions.

Fig. 3.3. Street-tree corridor. Credit: Pixabay/CC0 Public Domain.

Street trees

Street canyons are the narrow spaces between tall buildings that are often lined with trees (Fig. 3.3). The trees affect conditions by directing wind flow and lowering temperatures at ground level. Along roads and highways parallel to the wind direction, wind velocity in street canyons increases. But trees along wind routes and in green spaces can help reduce wind speeds. Green or vegetation-lined corridors improve thermal comfort for pedestrians by providing shade and releasing water vapor through tree leaf transpiration. Because of their limited root system along streets, regular pruning and upkeep are necessary to maintain the health of street trees and ensure their effectiveness. Tree-lined streets across a city can affect the UHI intensity and surface temperatures at street level. This mitigation measure can reduce the overall influence of a large UHI dome.

Rainfall

Measurable rainfall in cities is shed from hard surfaces and removed through drainpipes and storm sewers. Urban landscapes are usually developed from agricultural or natural land, and the initial construction involves reshaping the existing topography. Once a city has been developed, flood peaks in streams and rivers often increase two to four times in comparison with pre-urbanization. Increases are due to street pavements and roadways that cover surfaces, which reduces the potential for rainwater to infiltrate soil, which in turn increases runoff.

Warmer oceans due to climate change increase the amount of water that evaporates into the air. When more moisture-laden air moves over land or converges into a storm system, it produces heavier rain and snow storms. In recent years, a larger percentage of

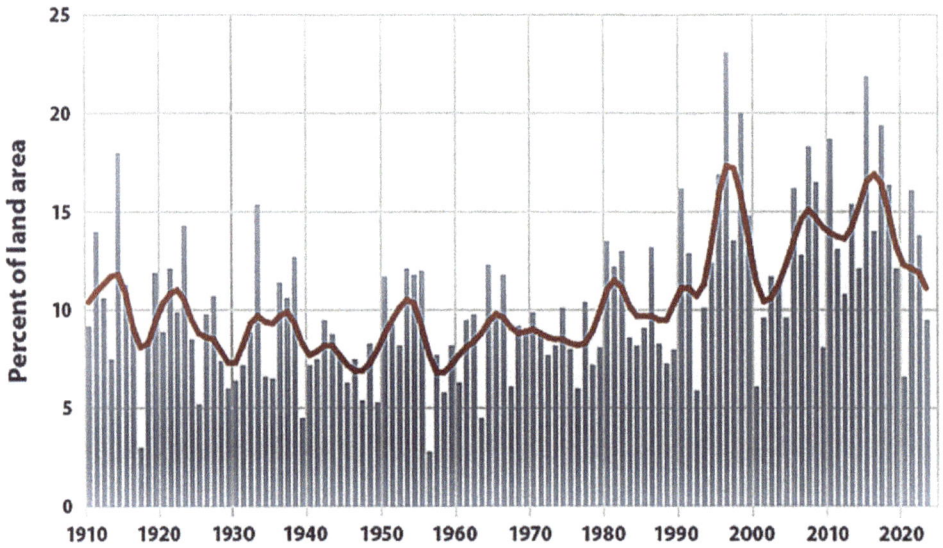

Fig. 3.4. Extreme one-day rainfall events in USA. Red line is 9-year weighted average. Credit: Chu *et al.* 2023.

precipitation has been in the form of intense single-day events (Fig. 3.4). The "sponge city" concept is designed to be a response to excessive rainfall and related weather events. It is a water management program to mitigate excess soil water resulting from extreme rainfall. It is based on modifying infrastructure to absorb, store, and purify rainwater. The criteria for a sponge city are an increased number of trees, lakes, and ponds to absorb rainwater. Also included is a planned vegetation canopy to redirect runoff by intercepting rainfall. Vegetation root systems can reduce runoff and encourage absorption to increase soil moisture which can promote vegetation.

The dome of warm air that is regularly over large cities forces moisture-laden clouds upward into colder air above, which initiates rain. Solid, liquid, and gaseous contaminants characterize the air of most modern cities. About 80% of the solid contaminants are particles small enough to remain suspended in the air for long periods. These particles directly influence rainfall and the air temperature in cities because the particulate matter provides nuclei for the condensation of atmospheric moisture into rain drops. The

general rule is, as cities increase in size, air pollution increases, and rainfall increases.

Urban air pollution island

Pollutants are generated from typical urban features and activities. The difference between urban areas and suburban and rural areas defines the intensity of urban pollutants (Fig. 3.5). They are not likely generated from sparsely populated areas where there is less industry and vehicle traffic. Urban heat islands and air pollution are not independent phenomena. Where urban heat islands exist, most likely urban pollution islands coexist. Research shows that 20 of the 24 global megacities have air pollution concentrations at levels where health effects are reported.

The pollution dome is another atmospheric condition caused by air pollution. It occurs when pockets of stagnant air form due to high levels of air pollution. It develops when pockets of air become trapped by the landscape, preventing air from moving. The stagnant air then becomes a dome that traps pollutants. These domes form at night when the air is not moving, and

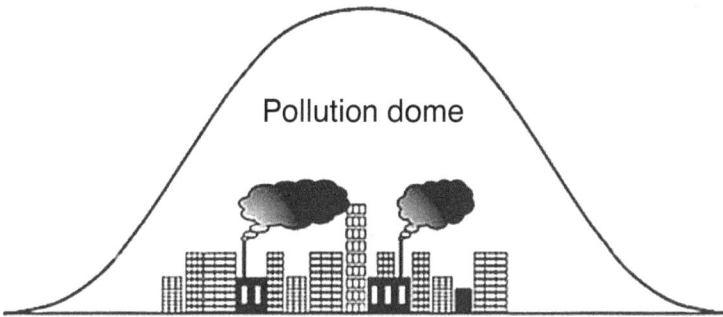

Fig. 3.5. Urban air pollution dome. Credit: W. Robinson.

Pollution dome

Fig. 3.6. Urban wind deflection patterns. Credit: W. Robinson.

particulate matter and gases are trapped in the still air. Pollution domes and urban heat islands are closely related. Pollution domes trap hot air within the dome and create an area of intense heat, and this trapped hot air contributes to the urban heat island effect.

Urban dry island

The "dry island" concept links the reduced humidity of urban areas to heat stress experienced by residents. Urban residents suffer more heat burden than the general (non-urban) population due to the urban heat island phenomenon. However, this exposure omits the influence of the urban microclimate phenomenon, the urban dry island. This concept considers that urban land tends to be less humid than the surrounding rural land. In dry, temperate northern climates, urban residents may be less

heat stressed than rural residents. However, in humid southern climates, the urban heat island is dominant over the urban dry island, resulting in 2–6 extra heat-stress days per summer.

Wind

Prevailing winds decrease speed over the landscape of towns and cities. Wind velocity may be half what it is in the open countryside. One reason for this is the surface texture caused by short and tall buildings (Fig. 3.6). Along roads and highways parallel to the wind direction, wind velocity increases, but trees along wind routes and in green spaces help reduce wind speeds. When wind reaches the wall of an urban building, it gets deflected in all directions. Some is deflected upward and around the sides of the building. A portion is deflected downward along the building wall, causing turbulence at ground level. The downdraught effect becomes more severe as the height of a building increases.

Fig. 3.7. Urban "cold island" effect. Credit: Pixabay/CC0 Public Domain.

Urban cold island

Urban green and blue spaces can produce "cold island" effects on the city landscape (Fig. 3.7). In some locations they can reduce the urban heat island effect by increasing shade and humidity which contributes small patches of cool air to the surroundings. Temperature reduction can be as much as 4°C in the affected areas. However, the effect on cooling and humidifying does not increase with area. The effect of a group of small green spaces can do more than a large area. Vegetation can lower air temperature via water evaporation, but it can sometimes increase heat burden because of humidity; this humidifying effect may erase the cooling benefit.

Urban light island

The "light island" refers to artificial light at night (ALAN). This condition alters the natural patterns of light and dark in urban areas. ALAN is increasing globally, with an estimated 80% of the world's population currently living under a "lit sky" (Fig. 3.8). And the amount of artificial light on the Earth's surface is increasing by at least 2% annually. In urban areas, artificial light comes from many sources and light levels may vary during the night. Street lighting is usually maintained at a constant level throughout the night, but light from the windows of residential buildings may only be important during the first hours of the night. Other sources of light pollution include industrial, institutional, and commercial sources, such as shopping centers, hotels, and vehicles.

About 30% of vertebrates and more than 60% of invertebrates are nocturnal. These species are most vulnerable to ALAN. Global surveys of night sky brightness show that 23% of land surfaces between 75°N and 60°S, including 88% of Europe and 47% of the USA, experience "light pollution" in the form of an 8% increase above natural level in night sky brightness. This level of light at night is likely to disrupt populations of crepuscular and nocturnal animal species.

Street lights

Urban street lights and commercial outdoor lighting (Fig. 3.9) have contributed to the presence, pest status, and probably the geographic distribution of some arthropods in the urban environment. A variety of flying insects are attracted to bright lights at night. This behavior often results

Fig. 3.8. Urban "light island" effect. Credit: Pixabay/CC0 Public Domain.

Fig. 3.9. Artificial light at night (ALAN) in Europe. Credit: E.esa.int/ESA.

in insects indoors and outdoors at windows or screens, and dead and dying insects near the light source. Factors that influence whether insects fly to outdoor lights include brightness (wattage), their ultraviolet output, and the heat produced.

Insects attracted to ALAN often exhibit spiraling flight patterns, while others approach directly. Some simply orbit the light source, frequently changing their flight speed and direction to remain within its vicinity. About 30–40% of insects that approach street lamps die soon after because of

collision, overheating, dehydration, or predation. ALAN may impede movement among habitat patches, lure individuals into bodies of water, or divert them into traffic. Insects not killed may become trapped in a "light sink," unable to forage, search for mates, or reproduce, especially when sexes differ in their attraction to light. Road surfaces, especially those dark colored, exposed to daylight can act as heat sinks. At night, insects and other arthropods orient to the warm road surface or fly in the air above it.

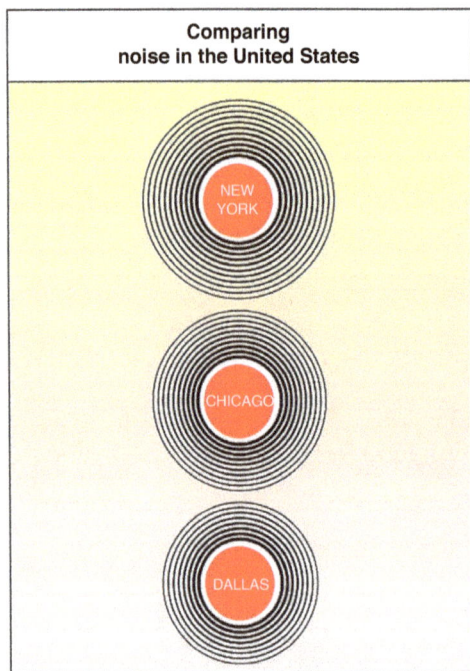

Fig. 3.10. Street noise (dB) in New York, Chicago, and Dallas, Texas, USA. Credit: Bartz Jul 2017 Bazaar.

Fig. 3.11. Food waste in an urban park. Credit: Pixabay/CC0 Public Domain.

Noise and Food Waste

Humans, particularly in large numbers and in cities, are noisy. Environmental noise is an increasingly common feature of urban areas and is described as an unwanted or undesirable sound within non-occupational settings. The most important contributor to urban environment noise is street-level sound. Another common feature of urban areas is discarded human food waste on streets and in parks. This organic material can become a reliable food resource for populations of squirrels, brown rats, gulls, and crows.

Street noise

The mean urban street noise level is approximately 73.4 dB, with substantial variation (range 55.8–95.0 dB) linked to conditions (Fig. 3.10). The density of vehicle traffic is associated with excessive street-level noise. The noise level of cities due to vehicle traffic may be higher during off-peak hours than during rush hours, probably due to the faster traffic speed during these hours. Urban sites generate and reflect high levels of anthropogenic noise during the day and night. Nightly noise levels can range between 30 and 60 dB, while levels in natural environments are usually below 30 dB. A noise level ≥ 70 dB has significant implications for human health, including links to sleep disorders, learning impairment, and heart disease.

Bats avoid noisy environments because the noise affects their ability to forage, communicate, and spatially orient themselves. Although urban environments constantly produce noise, it is unlikely that this noise would impact bat abundance and diversity. Much of the noise in urban environments, including that produced by transportation, is at relatively low frequency. Most bats are sensitive to frequencies above 10 kHz, as these are used in echolocation. In contrast to their response to rain noise, bats do not delay nightly roost emergence in response to traffic noise. Bats can discriminate between traffic noise and rain noise.

Food waste

The success of many urban animals is directly linked to their diet of processed food and human food waste (Fig. 3.11). Flocks of pigeons, gulls, crows, and starlings are a feature of major cities worldwide. The flocks of pigeons in city centers, crows vandalizing curbside garbage

Atlanta, GA   **Houston, TX**   Land surface temperature (2020)

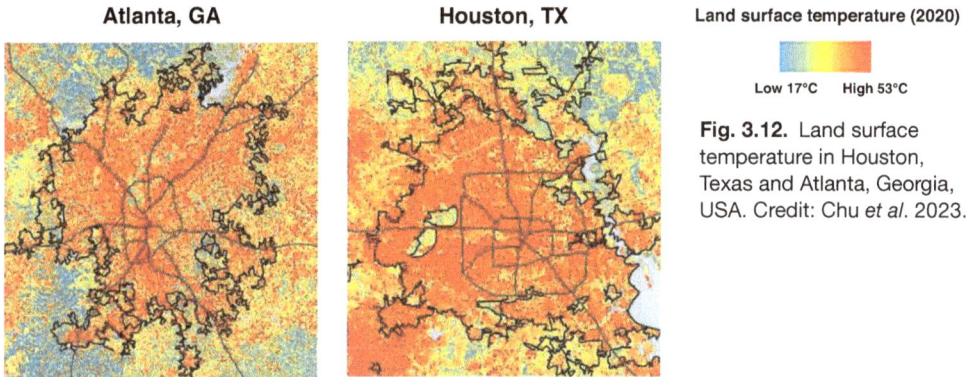

Low 17°C   High 53°C

Fig. 3.12. Land surface temperature in Houston, Texas and Atlanta, Georgia, USA. Credit: Chu *et al.* 2023.

containers, or gulls stealing food from tables have been an integral part of cities for many years. These birds are well adapted to breading, feeding, and nesting in the urban biome, and they establish large populations there. The external surfaces of urban buildings, roofs, fire escapes, and bridges provide roosting and nesting sites. Their food source is the system of landfills that contain the organic waste of the metropolitan area. This is their primary food source, while nesting and additional feeding is done in the city.

Land Surface Temperature

The land surface temperature (LST) measures how much the incoming solar energy interacts with and heats the ground, landscape features, such as green and blues spaces, and the surface of the canopy in vegetated areas. It is directly linked to local climate and influences local conditions and the ecosystems affected by those conditions. The impact of LST on large cities can be significant in terms of warm season temperature extremes.

The impact of the tree canopy on the land surface temperature in urban areas may be seen in the surface temperature of Houston, Texas, and Atlanta, Georgia, USA (Fig. 3.12). The tree canopy coverage for Atlanta is significant, with approximately 48% of the city's land area covered by trees. By contrast, Houston has approximately 18% tree canopy cover, but the majority of these trees are on private land.

Houston is one of the top five cities with the most intense urban heat islands. City infrastructure and local climate can influence land surface conditions, but tree canopy coverage can usually be increased by city planning.

Building surfaces and roadways act as heat sinks when they retain solar heat gained during the day and remain above ambient air and surface temperature at night. Green spaces and blue spaces are also affected by incoming energy influence on turf and aquatic ecosystems in these spaces. Sunlight impacts water temperature, oxygen levels, and the growth of aquatic plants. But excessive sunlight and solar energy result in high water temperature, reduced oxygen, and excessive algae growth. Reduced oxygen can stress aquatic organisms and alter community structure and the types of species that can survive in these conditions.

In many cities, the factors of rising temperatures due to the urban heat island effect and heat-absorbing pavement have changed the cycle of trees budding and shedding their leaves. In northern regions, artificial light is showing a bigger role than heat in changing plant cycles. Heat and light together seem to be causing trees to start growing in northern cities approximately 13 days earlier and end the season 11 days later. The additional growth time may not be beneficial, as it can add water balance strain for trees in cities experiencing reduced precipitation.

Elevated nighttime temperatures would fit the feeding behavior of some insect species, such as house mosquitoes (*Culex* spp.), which typically feed at night, and the American cockroach, which forages away from the sewer system and readily flies during warm nights. Native animal and plant species in green medians of highways may have an extended growing season as a benefit of the moderated temperatures in spring and fall. Invasive plant and animal species using the medians and roadway edges as dispersal routes are likely to benefit from the temperature conditions. Building surfaces that retain solar heat into the night frequently attract insects and other arthropods to the edges of doors and windows.

Additional Reading

Brügger, L., Lardon, J., Müller, G. and Hufschmid, T. (2014) Tick abundance in the city of Zurich, Switzerland. In: Müller, G., Pospischil, R. and Robinson, W.H. (eds) *Proceedings of the Eighth International Conference on Urban Pests*, ICUP, Zurich, Switzerland, p. 425. Available at: https://icu p.org.uk/media/hynpdvrq/icup1154.pdf (accessed 29 September 2025).

Caldirola, D., D'amicis, C. and Massara, P. (2025) Pests associated with green buildings: An emerging infestation in Italy. In: Robinson, W.H (ed.) *Proceedings of the Eleventh International Conference on Urban Pests*, ICUP, Lund, Sweden, pp. 125–130.

Dorn, W., Messutat, S., Kipp, S., Sûnder, U., Feldman, A. *et al.* (1999) Seasonal variation in the infestation of rodents with *Ixodes ricinus* (Acari: Ixodidae) and prevalence of infection with *Borrelia burgdorferi* in a recreation area. In: Robinson, W.H., Rettich, F. and Rambo, G.W. (eds) *Proceedings of the Third International Conference on Urban Pests*, ICUP, Prague, Czech Republic, pp. 463–469. Available at: https://icup.org.uk/media/cpuhemi3/icup459.pdf (accessed 29 September 2025).

Hirabayashi, K. and Zukeran, H. (2017) Increase in environmental light conditions boosts massive flights of aquatic insects. In: Davies, M.P., Pfeiffer, C. and Robinson, W.H. (eds) *Proceedings of the Ninth International Conference on Urban Pests*, ICUP, Birmingham, UK, pp. 53–59. Available at: https://ic up.org.uk/media/ygvj2c5v/icup1182.pdf (accessed 29 September 2025).

Inoue, E., Kimura, G. and Hirabayashi, K. (2008) Chronomids (Diptera: Chironomidae) attracted to vending machines in the middle reaches of the Shinano River, Japan. In: Robinson, W.H. and Bajomi, D. (eds) *Proceedings of the Sixth International Conference on Urban Pests*, ICUP, Budapest, Hungary, pp. 177–185. Available at: https://icup.org.uk/media/3hbnmhzi/icup900.pdf (accessed 29 September 2025).

Lozzia, G.C., Ottoboni, F. and Rigamonti, I.E. (1993) Possibilities for integrated pest management on urban trees. In: Wildey, K.B. and Robinson, W.H. (eds) *Proceedings of the First International Conference on Urban Pests*, ICUP, Cambridge, UK, p. 488. Available at: https://icup.org.uk/media/wo5pohqc/icup6 87.pdf (accessed 29 September 2025).

Mattila, M. and Burgin, S. (2014) Queensland's newest invasion: Feral urban deer. In: Müller, G., Pospischil, R. and Robinson, W.H. (eds) *Proceedings of the Eighth International Conference on Urban Pests*, ICUP, Zurich, Switzerland, pp. 297–302. Available at: https://icup.org.uk/media/td3ldatq/icup1126 .pdf (accessed 29 September 2025).

Medeiros-De-Sousa, A.R., Vendrami, D.P., Wilke, A.B.B., Urbinatti, P.R., Ceretti-Júnior, W. *et al.* (2014) Mosquito assemblage diversity in urban parks, São Paulo, Brazil. In: Müller, G., Pospischil, R. and Robinson, W.H. (eds) *Proceedings of the Eighth International Conference on Urban Pests*, ICUP, Zurich, Switzerland, pp. 147–152. Available at: https://icup.org.uk/media/t5ijo0n4/icup1103.pdf (accessed 29 September 2025).

Wang, L., Meng, L., Richardson, A.D., Hölker, F., Li, H. *et al.* (2025) Artificial light at night outweighs temperature in lengthening urban growing seasons. *Nature Cities* 2, 506–517. DOI: 10.1038/ s44284-025-00258-2.

4 Ecological Fitness

The first observed adjustment by an animal species to the urban biome is usually a behavior modification. Many natural species change their foraging and feeding habits and diet composition in response to the food available in this new habitat. Sometimes the change is to utilize human food (Fig. 4.1). This difference in food selection between species in urban and natural environments may simply indicate the coupling of a behavior trait with an abundant food resource. In urban environments, various species have adjusted their behavior to feeding on exposed human food. In terms of abundance, the availability of human food is matched by populations that are dependent on it. The ecological trap set for some of these species is that the population size may become larger than the resource can support. The trap may close when human food waste is better managed and less available as a food resource.

Food Waste and Nutrition

Ecological traps develop when a preferred habitat or a critical resource has negative fitness consequences or is abruptly changed by human actions. In the case of bird species that are considered food scavengers, routinely feeding on human food waste might not provide adequate nutrients or might be detrimental. A diet based primarily of human food could have negative consequences during the breeding season, when parents are foraging food waste and feeding it to their young. Human foods tend to be carbohydrate rich but protein poor. This may benefit males because they are larger and may generally have an increased calorie intake. But females, particularly reproducing females, often have limited foraging time because of their nesting habits. They may experience fewer benefits from human food waste.

For some urban species, artificially increased food abundance in the form of human food waste might be a false indication of food quality and create an ecological or nutritional trap. The expectation for bird species regularly eating human food waste is that it is detrimental to the overall health of adults and nestlings. A study that examined the effect of human food on the cholesterol level of crows found that nestlings fed a diet supplemented with fast food had higher cholesterol levels than those fed un-supplemented food. However, the body condition and 2–3 year survival of the nestlings indicated cholesterol levels had no detectable effect on survival. Instead, they were correlated with higher indices of body condition. But the potential long-term negative effects may include resistance to disease and the gradual loss of natural foraging behaviors.

Fig. 4.1. Grey squirrel. Credit: Pixabay/CC0 Public Domain.

Urban Landfills

The continuous increase in the global population means human waste generation is also on the rise. Early waste disposal practices primarily involved methods such as open dumping and burning. The establishment of modern landfills for household waste began during the mid-20th century. These sites were generally small and designed and designated to contain general waste materials only from the surrounding community. Their design has been improved and the environmental impact reduced. Modern waste management is typically made up of one or more large landfills and a system of transfer stations (Fig. 4.2) at the regional level. Transfer stations are where municipal waste is transferred from curbside collection trucks to large vehicles for transport to a landfill. Transfer stations are partially open structures and the exposed organic waste usually attracts birds and other foraging animals.

Landfills will likely continue to be an important endpoint of waste management for large metropolitan and suburban areas. However, the amount, processing methods, and use of the organic portion of waste will change. The amount will decrease as the next generation of households and food facilities become more sustainable-minded and reduce or redirect what can be used or composted. This change will develop slowly but will likely affect the major bird pests in cities. The life cycle of starlings, gulls, and pigeons that is based on the dual habit of foraging at transfer stations and landfills but nesting in cities, may realign the urban populations of these species. Human food will always be available at public locations, but the demand for it from feral birds may decrease and their urban populations decline.

Urban Temperatures

Dissipating excess heat built up during days of elevated temperatures has survival value for birds, especially species in urban environments. Birds rely on panting, gular fluttering, and cutaneous heat loss for evaporative heat dissipation. Access to water in urban areas can lower the risk of dehydration during extreme heat events. But exposed water sources increase the risk of breeding sites for disease-carrying mosquitoes. Availability of water for urban bird species improves their water reserves, which are needed for evaporative cooling and to maintain body temperature below lethal limits. Even with access to water, they may succumb to hyperthermia if they are not able to dissipate heat during extremely hot conditions. During hot spells, many birds reduce foraging activity to minimize heat gain.

Fig. 4.2. Waste collection, transfer station. Credit: Pixabay/CC0 Public Domain.

Fig. 4.3. Urban pigeon color patterns. Credit: Pixabay/CC0 Public Domain.

Fig. 4.4. Urban pigeon. Credit: Pixabay/CC0 Public Domain.

This can result in a costly trade-off, because it can affect their ability to maintain body mass and continue nesting and feeding young.

Urban heat islands and surface temperatures influence the behavior of most urban species. For example, feral pigeons exhibit behavior that is adaptable to extreme urban conditions. This results in their distribution across moderate and cold climatic zones, especially in urban areas where food and water are readily available. But pigeons also inhabit dry, hot deserts in Asia and North Africa. In urban settings, the degree of urbanization influences pigeon density, distribution, and activity patterns. Coloration patterns are linked to a thermoregulatory capacity and energy expenditure in response to habitat temperature (Fig. 4.3).

Pigeons

This common synanthropic species lives in both urban and agricultural environments worldwide. They are gregarious animals that gather in large flocks at resource-rich locations. They utilize existing conditions and are capable of roosting on building surfaces in cities (Fig. 4.4). The success of pigeons in urban settings is attributed to the availability of building ledges and roof overhangs that simulate natural spaces of their original habitats, such as cliff faces. They can avoid

or tolerate the heat stress of some urban heat domes by adjusting daily activity. Pigeon abundance in cities can also be linked to foods rich in carbohydrates and protein, in the form of scattered food waste or food routinely supplied by humans in urban parks and open spaces.

The pest status of these large birds is generally based on numbers of individuals in public spaces, and the damage done by concentrations of feces. Examples of high urban densities include 5 birds per hectare in inner-city Amsterdam (Netherlands); 6.8 per hectare in Wellington (New Zealand); 8.1 per hectare in Padua (Italy); 9.5 per hectare in Barcelona (Spain); and up to 20.8 per hectare in inner city Milan (Italy). The density appears to be directly related to the amount of food, including intentional feeding, spillage, and organic waste, which is then related to the size of human population. A similar association with availability of food occurs with brown rat sightings in large-population cities.

Feral pigeons possess several notable urban survival traits, including broad acceptance of food types and sociability, that support establishment in new environments. They readily investigate and sample new food resources and consume a variety of food types. Pigeons in cities often feed largely on discarded human foods. While these foods do not bear much resemblance to natural foods for this species, such as seeds, they are readily available year-round. The feeding behavior of older, experienced birds seems to be the main factor that induces inexperienced young pigeons to eat new foods. Feeding at landfills likely provides a variety of new foods, and feeding preferences may originate there.

Colonies usually form near consistent food resources, although some members of urban colonies travel daily to food resources outside the city. Since pigeons can store a large quantity of food in their crops and need only a few minutes to fill the crop, they can use food sources spread widely across urban habitats. Because of their tendency to visit many foraging sites each day, individual pigeons can monitor potential food sources and return to them when they become consistent. Abundant, accessible food may be a prerequisite for colony formation and persistence, while solitary nesting habits may be the result of food availability stress.

Pigeon excreta is a major problem due to the soiling of building facades, internal spaces, and outdoor statues. Their feces can be harmful to humans and domestic animals because it contains salts: phosphoric, nitric, and uric acids. Acid deposited on surfaces reacts with the components and corrodes or dissolves the surface. Uric acid will cause the decay of structural and ornamental sandstone, limestone, and marble, metals in outdoor sculpture, and paint finishes (Fig. 4.5).

European Starling

This generalist species thrives in a wide variety of environments, particularly those altered by humans. Starlings are considered as "urban exploiters" due to their ability to adapt to new conditions and establish large populations in urban environments (Fig. 4.6). They are generalist consumers and compete with native species for food and often take the nest sites of other birds. They were introduced into North America in 1890 and rapidly adjusted to conditions not found in their native European range. The changes happened in a span of just 130 years. The total continental population reached an estimated high of 200 million starlings in the 20th century.

While starlings are successful in the urban biome, they prefer cleared agricultural and suburban areas. This preference is evident in their abundance and can be linked to production agriculture. Since 1964, estimated starling numbers in Great Britain have declined by more than 50%. Finnish starlings declined by 90% from 1970 to 1985, which corresponded to a decrease in cattle farming across the country. A shift to indoor cattle husbandry in Denmark may have contributed to the 60% decline in starling abundance between 1976 and 2015. Changes may be related to recent population declines in areas that were once rural and have become more urban.

Fig. 4.5. Pigeons roosting on statue. Credit: Pixabay/CC0 Public Domain.

Fig. 4.6. European starling. Credit: Pixabay/CC0 Public Domain.

Pest status is based on density and their acceptance of a range of food types. They forage in large flocks and mostly on the ground in open areas, whether farm of urban park. Their feeding habits in urban areas include human food waste and discarded scraps, and they often scavenge in open trash bins in cities. Along with pigeons, crows, and gulls, they can occur in large number on urban landfills and at municipal waste transfer stations. Commonly, as much as 50% of all the birds at transfer stations are European starlings. They use these buildings for nesting and daytime resting sites and forage there on food waste. Transfer stations and open or working landfill sites are an integral part of the foraging routine for starlings.

Starlings have several natural predators, especially during breeding, nesting, and roosting cycles. These include species of small hawks, falcons, owls, and peregrine falcons. They frequently disperse from roosting sites in large numbers, and these flocks are called murmurations. This concentration of individuals apparently provides some protection, as predators have difficulty targeting one bird in a flock of thousands, and the flock can rapidly change direction (Fig. 4.7). Recent studies using models of animal movement show these patterns emerge naturally when birds coordinating their movement with a number of nearby neighbors.

Crows

Crows, magpies, ravens, and jays (Family Corvidae) have been successful in adjusting to urbanized environments around the world. At least 23% of the 130 corvid species live in urbanized environments. Factors contributing to the success of these birds in the urban environment include food availability and nesting sites (Fig. 4.8). Breeding in urban environments is associated with early nesting, high fledging success, reduced home range size, and limited territoriality. Despite geographic location, corvids show flexibility in resource use that enables them to exploit novel substrates for nesting and feeding.

Corvids as a group, and especially crows, are highly synanthropic and easily adjust to modified urban environments. The primary traits of corvids that enable them to exploit new, urban environments are their high behavioral plasticity and flexible resource use. Easily accessible food is the most important resource attracting these birds into cities. Crows display similar behavioral adjustments in different cities around the world. egg

Fig. 4.7. European starling murmuration. Credit: Pixabay/CC0 Public Domain.

Fig. 4.8. Crow. Credit: Pixabay/CC0 Public Domain.

Fig. 4.9. Crow eating food waste. Credit: Pixabay/ CC0 Public Domain.

laying starts earlier in urban populations than in non-urban populations, and there are often larger groups or clutches in urban than rural populations. They are successful urban exploiters of the modifications of ecosystems and infrastructure caused by urbanization, for instance using buildings, poles, and power lines as nesting sites. Flocks in rural areas and suburban areas can include ten or more individuals that roost and forage together.

Crows in cities eat mostly human food waste at refuse containers near buildings and along city streets (Fig. 4.9). In some cities the timing of the refuse collection and crow feeding coincide in favor of the crow. A human refuse meal would include meat, bread, and vegetables, and this would account for about 65% of a crow's diet in urban areas. In the USA, where the average household

wastes nearly 40% of its food supply, crows are capable of thriving on this single resource.

The pest status of crows in urban areas is based on foraging human food waste in refuse containers and the increase in crow numbers. The urban crow population in Tokyo began increasing in the 1980s and peaked around 2000. In 1985, research teams counted 6737 crows at locations around Tokyo, and by 1990 this figure increased to 10,863. The increase was linked to the amount of garbage generated and mishandled in the city. About 3.97 million tons of waste were collected in 1985 and this increased to about 4.8 million tons in 1990. A decrease in the amount garbage created, along with the management of exposed garbage, influenced a significant drop in Tokyo crow populations. About 3.52 million tons of food waste were generated in 2001, which dropped to about 2.55 million tons in 2020. In September 2001, a research team estimated that about 36,400 crows were living in Tokyo, but the number diminished to about 11,000 in 2020.

Herring Gull

Gulls are unusual among seabirds in that several species in various parts of the world readily breed and forage in urban areas. Populations of these species have increased throughout coastal areas of North America and Europe. The increase is attributed to protection from human disturbance, availability of human food waste, and the ability to adjust to human-altered environments. The herring gull is the best known of the large gull species (Fig. 4.10). It is a white bird that nests in colonies ranging from a few individuals to hundreds of pairs.

Overfishing and climate change influence on ocean water temperatures have caused some fish species to shift their distribution. This decrease in food availability at previous foraging sites resulted in herring gulls shifting their distribution into urban areas. There seems to be no shortage of breeding sites and human food refuse in coastal cities and towns. In these locations, gulls benefit from shorter foraging distances, obtaining more food with each trip, which promotes successful breeding. Nesting sites and substrates on buildings include chimney stacks, roofs, ledges, pipes, and vents. Gulls prefer to nest on buildings that are close to streetlights and food sources. They forage during the day, but urban street lighting may extend foraging activity to local food sites that are also illuminated at night.

Herring gull foraging trips to urban food habitats and nest sites on Long Island, New York, USA, demonstrate fidelity to a site in terms of the food sources revisited during trips (Fig. 4.11). This foraging behavior indicates the survival value of using predictable sources such as dumpsters and landfills. The adults foraged at and revisited these sites frequently, often had similar fight paths, and regularly visited sites in the same order on each trip. Urban gull populations primarily consume human food waste. Urban gulls tend to lay eggs slightly earlier than females in rural locations. This may be linked to their diet or the urban heat island effect. Urban populations have less predation and higher fledgling success than rural counterparts.

The foraging habits of herring gulls are highly dependent on the abundance and distribution of human sources of food. Population expansion of species in some regions may be the direct result of increased availability of human food waste. The reliance of herring gulls on urban food resources could make them more vulnerable to changes in human behavior regarding the handling of waste. Sustainability measures regarding household waste and urban landfills may change and

Fig. 4.10. Herring gull. Credit: Pixabay/CC0 Public Domain.

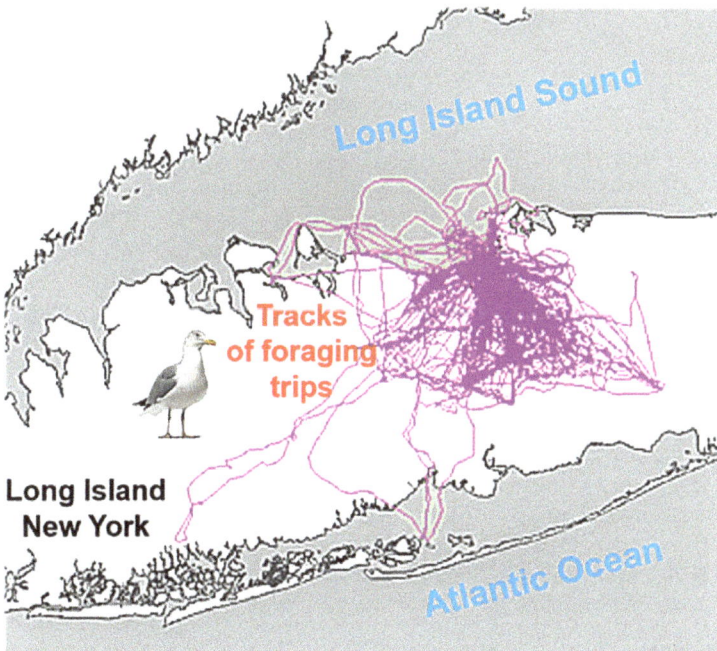

Fig. 4.11. GPS tracks (color) of foraging trips by herring gulls from landfill site to urban habitats. Credit: Adapted from K. Lato *et al.* 2021.

Fig. 4.12. Monk parakeet. Credit: Pixabay/CC0 Public Domain.

Fig. 4.13. Monk parakeet nest. Credit: Pixabay/ CC0 Public Domain.

alter the food available. This could be significant and create an ecological trap for urban gull populations that over-depend on this food resource and have a reduced ability to return to natural foraging behaviors.

Monk Parakeet

The monk parakeet is one of the most invasive bird species (Fig. 4.12). Currently, it is considered an invasive species in 19 countries, but this may not be the end of its dispersal. The global distribution of this parrot is in part due to the pet trade market and hobbyists. On the positive side, in the urban environment the monk parakeet can contribute to the breeding success of other bird species. Monk parakeets build their own communal nests and make breeding space available for use by other species. In some cases, nest occupancy may include nine bird species, including native and invasive species.

Invasive populations of monk parakeets can be found worldwide. The first records of it in the USA date from 1968 in Florida. Monk parakeet populations in the USA are essentially urbanized

and are a valued component of local avifauna in many communities. In southern Florida, they build nests principally on man-made structures such as stadium light poles, mobile phone transmitting towers, and electric utility facilities in urban areas (Fig. 4.13). They are considered a nuisance by utility companies due to their link to power failures. Since establishment in Florida, they have continued to expand their distribution through North America. In Europe, the first establishment was in Barcelona, Spain, in 1975, where populations are now estimated to double every 9 years.

The breeding season of the monk parakeet is restricted to spring and summer and apparently controlled by photoperiod. Changes in this pattern may be related to environmental factors such as temperature and food supply. The clutch size is larger than that of most parrots, which may compensate for a generally low fledging and breeding success. A lower breeding success may be based on the elaborate "stick nest" built in urban areas. Although a closed stick nest is safer than an open nest, it is less secure on some surfaces than those made in tree or cliff cavities. This may have favored selection for a faster growth rate of nestlings and a shorter fledging period.

Fig. 4.14. *Psitticimex uritui*, a parasite of monk parakeets. Credit: Iorio *et al.*, 2010.

Fig. 4.15. Eastern grey squirrel. Credit: Pixabay/ CC0 Public Domain.

In old nests the blood-feeding cimicid *Psitticimex uritui* (Fig. 4.14) becomes abundant. The presence of this parasite may be a disadvantage for long-term occupation of large nests. During breeding season, monk parakeets typically add green vegetable material to the lining of their nests. This behavior may help to limit the presence of the cimicid parasite. Low nest site fidelity may be related to nest infestations; some populations can have an average of 47% of the mating pairs changing nest sites between years.

Squirrels

The squirrels in urban habitats are food generalists and opportunists. They change feeding behavior according to habitat and environmental conditions. Despite favoring seeds and nuts, their diet depends on human food sources, which enables their success in urban habitats. But urban habitats can act as ecological traps, if feeding on human food items results in an imbalanced diet. Effects of a poor diet are reduced physical ability and decreased disease resistance.

Eastern grey squirrel

This species has some white and brown features (Fig. 4.15) and is a common inhabitant of wooded urban and suburban parks in its native and introduced range. Urban populations can occur at much higher densities than their rural counterparts. Human-based food sources have been suggested as the main reason for elevated populations. The density in urban areas can be up to 100 times that of rural areas. Urban habitat variables, such as vegetation, may be less important drivers of density if human food, which can comprise 35% or more of the diet, is available. The black morph of this species increases with the extent of urban land cover. It is a successful morph in both urban and suburban sites, and populations can be solely this color.

Grey squirrels are linked to frequent urban and rural power failures. They possess sharp claws and a strong grip which allows them to balance on small-diameter power cables. Gray squirrels frequently use electrical power and telephone wires as travel lanes and move rapidly along these wires. When a squirrel climbs on an electrical transformer, it may cross the high-voltage wire and trigger a power outage (Fig. 4.16). The American Public Power Association uses "The Squirrel Index" (TSqI) to track squirrel attacks on electrical power systems. The TSqI Index identifies two peak periods of disruption: May–June and October–November.

European red squirrel

This species is declining in its European forest habitat, and its occurrence in urban areas may indicate an important alternative habitat. The loss of forested land due to urbanization may

Fig. 4.16. Electric power transformer. Credit: Pixabay/CC0 Public Domain.

Fig. 4.17. Little brown bat adult. Credit: Pixabay/CC0 Public Domain.

be detrimental to this squirrel. Cities provide human food resources, nest sites, and few natural predators. When natural food sources are available in sufficient amounts, this squirrel uses human food only as a supplement. Whether urban parks provide adequate refuges for the European red squirrel might depend on the surrounding habitat and availability of natural food and shelter. Squirrel abundance is higher in urban and rural areas than in forest habitats.

Fox squirrel

This is a common species throughout the Midwest and southern USA. It has adjusted to urban areas by utilizing large trees as preferred nest sites. Females select buildings that provide a warm refugia during winter and spring to raise young. The use of buildings is more pronounced in northern urban areas where the number of natural nest cavities is a limiting factor. Fox squirrels spend considerably more time on the ground than other tree squirrels. Their ability to tolerate paved surfaces during daily movements and to use buildings as shelter sites opens urban environments to them.

Bats

Urbanization favors some bat species, which have succeeded in human-altered conditions. They likely benefit from the increased temperatures of typical urban areas and insects at lights at night. Water for drinking, foraging, and ease of navigation are key habitat requirements for bats. Water may be more abundant in cities than the natural landscape because of the ponds and small lakes in urban green spaces, and residential swimming pools. The loss of natural roost sites in urban areas has been offset by sites available on building surfaces. These may simulate the structural properties found in cliffs, caves, or trees that common species use as roosts. Adjusting to the urban surfaces provides a warm microclimate for reproductive females, which may advance the timing of reproduction and improve growth and body size of young.

ALAN affects bats by delaying night emergence times, impeding movement, and reducing foraging activity. The effects on bats are often highly species-specific. Some species, such as the little brown bat (Fig. 4.17), display avoidance of lights, but others are attracted to them and the concentrations of insect prey. Light avoidance may be due to their slow flight speeds, which may increase their perceived predation risk in lit environments.

Urban environments generate and reflect high levels of noise during the day and night. Bats avoid noisy environments because the noise affects their ability to forage, communicate, and spatially orient themselves. Much of the noise in urban environments, including that produced by transportation, is relatively low frequency. Excessive noise can wake bats during hibernation, but species in urban environments generally acclimatize to their acoustic environment.

Bats may enter torpor, which is a state of reduced heart rate and respiration, for just a few hours to save energy, or they can remain in this state of hibernation for a month while overwintering. During hibernation, they cycle through brief periods of arousal when their body temperature returns to normal for a few hours. Some species, such as the little brown bat, may hibernate for more than 6 months waiting for the return of insects in the spring. White-nose syndrome (WNS) is a fungal disease that has killed millions of bats in North America. It is white fuzzy growth on the nose of infected bats. WNS repeatedly awakens bats from hibernation, causing them to consume their winter fat stores, which can result in starvation before spring.

The pest status of bats in general, and especially those common in suburban and urban areas, is based on perception. Folklore and misinformation have established bats as something to be feared. Bats sometimes accidentally fly into houses or enter from a chimney. Some species roost in attics or chimneys of houses or tall buildings. The benefits of bats to the ecological environment are rarely presented to the public.

Stray Dogs

The worldwide population of domestic dogs is estimated at approximately 700 million, with about 75% classified as free-roaming unowned strays and feral. Free-roaming dog abundance varies between countries and is related to an urban or rural habitat and human population density. Stray dogs are successful in urban environments because they form small social groups and generally utilize the abundance of human food waste and available shelter. Where free-roaming dogs exist in high densities, there are concerns because of the risks to public health, animal welfare, and wildlife. Dogs are responsible for transmitting over 300 zoonoses to humans, dog bite injuries, and the spread and maintenance of the rabies virus worldwide; they are a primary reservoir host of this virus.

Identification of dogs in the urban environment includes the following categories:

Fig. 4.18. Stray dog in urban area. Credit: Pixabay/CC0 Public Domain.

Family dogs have owners on whom they depend and, although they may be free to roam, their reproduction is supervised by humans. *Stray dogs* (Fig. 4.18) include dogs living in a human-dominated environment. This is a heterogeneous group: it includes dogs that still have a social bond with humans, possibly abandoned or born in human settings, and which tend to associate food with households. Behavioral observations reveal that stray dogs occasionally form groups for communal defense of a territory. The stability of these groups may be linked to long-term bonds among group members. *Feral dogs* include all dogs living in a wild and free state with no direct food or shelter intentionally supplied by humans. They show no human socialization and usually have some avoidance of humans.

The activity of small groups is greatest during early morning (7:00–10:00 hour) and late afternoon or early evening (17.00–21.00 hour). A third activity peak corresponding with long solitary excursions may occur after midnight. On a typical day, the dogs spend up to 18 hour sleeping, lying, or sitting inside the territorial boundaries. Urban stray dogs feed chiefly on garbage refuse and occasional handouts from local area residents. Garbage is an unevenly distributed food resource, and considerable time can be spent foraging. Human food or food waste is available to the stray dogs in the form of small amounts that are usually widely scattered throughout their home ranges. This distribution of food resources promotes foraging on an individual basis or in small groups.

Different groups or government agencies are often responsible for stray dog population management programs. The three main strategies are: culling, long-term sheltering, and fertility control. Culling is based on the episodic removal and killing of individuals for the purpose of population reduction. In some countries, sheltering free-roaming dogs is the most common method of dog population control. The objective of sheltering is to reduce the free-roaming dog population size by removing dogs. Sheltered dogs may be euthanized or adopted or may permanently remain in the shelter. Fertility control is through surgical or chemical sterilization, or contraception. The strategy of catch-neuter-release of free-roaming dogs is the most common method of fertility control. It has positive effects on dog health and welfare, including improved body condition and reduced presence of injuries and some pathogens.

Fig. 4.19. Peregrine falcon. Credit: Pixabay/CC0 Public Domain.

Raptors

The raptors present in city centers are usually small species that are habitat generalists. They do well foraging and breeding in the variety of urban vegetation and terrain types. They nest in trees in urban green spaces, but also in artificial structures, such as on rooftops and in cavities in buildings. The fitness of small raptors to urban areas may be linked to the advantages of reduced metabolic needs provided by the urban heat island.

Fig. 4.20. Red-tailed hawk. Credit: Pixabay/CC0 Public Domain.

Peregrine falcon

These raptors have adapted to living in many cities and make use of ledges on tall buildings for nesting. Pigeons, several species of gulls, and starlings provide food for these predators. They also feed on rodents and ground squirrels (Fig. 4.19) when the conditions are right. Falcons primarily hunt in open areas, making them less effective in tight spaces or areas with obstacles.

The presence of this large raptor can create a significant psychological deterrent for pigeons, as they instinctively fear falcons. A mating pair of falcons can hunt over a large area and potentially displace pigeons from multiple locations and breeding sites. The pigeon is a key prey species for urban-dwelling raptors across the globe. They composed 32% of the peregrine falcon diet in Warsaw, Poland, with a peak of over 50% in the winter and summer. In southwest England, pigeons and doves make up 47% of the prey of peregrines. In New York City, pigeons can be 75% of the peregrine's diet.

Red-tailed hawk

This species (Fig. 4.20) is the most common hawk throughout the USA, Canada, Mexico, and Central America. They thrive in many habitats, including open fields, deserts, parks, woodlands, and forests. They usually occupy the same territory on a year-to-year basis.

These hawks hunt during the day and can be seen gliding over city parks and green spaces. Their excellent eyesight helps them spot prey in open areas. In the city center, hawks often perch on high platforms like telephone poles and wait for prey from above, then swoop down and capture them. They adapt their diet to whatever small animal is seasonally available. They consistently prey on rodents, rabbits, squirrels, and pigeons.

Predation by hawks on various prey species in cities has the advantage of captive populations and perhaps limited shelter from these predators. However, predation on other birds in flight may not be higher relative to what is hunted in woodlands. The presence of hawks can disperse pigeons and other birds from large sites, including airports and landfills. They provide pest bird control in urban areas where bird-proofing methods are difficult. Pigeons have an instinctive fear of hawks, and their presence can prevent pigeon populations from roosting or nesting and encourage adults to nest in alternative locations.

Additional Reading

Barrio, A.C., Cámara Vicario, J.M., García-Howlett, M., Serrano Casaus, M., Bueno Marí, R. et al. (2025) Study on urban rock dove (*Columba livia*) movements in the city of Madrid using GPS devices. In: Robinson, W.H (ed.) *Proceedings of the Eleventh International Conference on Urban Pests*, ICUP, Lund, Sweden. pp. 116–124.

Cámara, J.-M., Torres, P. and García-Howlett, M. (2014) Geographic information system in pest control programs, an example with feral pigeon (*Columba livia*) control program in Madrid city. In: Müller, G., Pospischil, R. and Robinson, W.H. (eds) *Proceedings of the Eighth International Conference on Urban Pests. ICUP*, ICUP, Zurich, Switzerland, p. 461. Available at: https://icup.org.uk/media/v52j03c1/icup1172.pdf (accessed 29 September 2025).

Cooper, D.S., Shultz, A.J., Şekercioğlu, Ç.H., Osborn, F.M. and Blumstein, D.T. (2022) Community science data suggest the most common raptors (Accipitridae) in urban centres are smaller, habitat-generalist species. *IBIS* 164, 771–784. DOI: 10.1111/ibi.13047.

Filho, H.O. (2011) Bats in natural and urban environments. In: Robinson, W.H. and Carvalho Campos, A.E. (eds) *Proceedings of the Seventh International Conference on Urban Pests*, ICUP, Ouro Preto, Brazil, pp. 13–14. Available at: https://icup.org.uk/media/p4ifi4kq/icup0957.pdf (accessed 29 September 2025).

Filho, H.O. (2011) Urban bats: Aspects of ecology and health. In: Robinson, W.H. and Carvalho Campos, A.E. (eds) *Proceedings of the Seventh International Conference on Urban Pests*, ICUP, Ouro Preto, Brazil, pp. 295–298. Available at: https://icup.org.uk/media/ilwf1h5h/icup0995.pdf (accessed 29 September 2025).

Lambert, M., Massei, G., Dendy, J. and Cowan, D. (2017) Towards practical application of emerging fertility control technologies for management of rose-ringed parakeets. In: Davies, M.P., Pfeiffer, C. and Robinson, W.H. (eds) *Proceedings of the Ninth International Conference on Urban Pests*, ICUP, Birmingham, UK, pp. 179–187. Available at: https://icup.org.uk/media/lzwn2vvp/icup1203.pdf (accessed 29 September 2025).

Pellizzari, M. and Loughlin, D. (2017) Controlling urban pigeon populations humanely. In: Davies, M.P., Pfeiffer, C. and Robinson, W.H. (eds) *Proceedings of the Ninth International Conference on Urban Pests*, ICUP, Birmingham, UK, pp. 171–177. Available at: https://icup.org.uk/media/42unjh30/icup1202.pdf (accessed 29 September 2025).

Shirai, M. (2022) Striking a balance between effort and benefit for bird damage control: Crow management in Tokyo, Japan. In: Bueno-Marí, R., Montalvo, T. and Robinson, W.H. (eds) *Proceedings of the Tenth International Conference on Urban Pests*, ICUP, pp. 364–368. Available at: https://icup.org.uk/media/ubudq2bb/70-shirai-167-f-pp-364-368.pdf (accessed 29 September 2025).

Smith, L., Hartmann, S., Munteanu, A., Dalla Villa, P., Quinnell, R.J. *et al.* (2019) The effectiveness of dog population management: A systematic review. *Animals* 9, 1020. DOI: 10.3390/ani9121020.

5 Harborage

Harborage is destiny for pests in the urban biome.

The importance of harborage to insects in the urban biome is profound. It is the functional connection between an organism and the habitat. The genesis of a species harborage is based on the opportunity to find eligible sites, and the frequency of those sites in the habitat. The habits of the German cockroach and the crevices in kitchen cabinetry were brought together in early cities. The introduction of indoor plumbing and central heating secured the German cockroach connection with the household habitat. The suitable crevices were conditioned by adults and nymphs with pheromones and feces for the long-term production of generations of conspecifics. These harborages were individual anchors of large populations that infested indoor habitats.

Harborage is the base of synanthropic populations in the urban biome. Populations are supported by harborages, and the conditions there provide security and influence group behavior, foraging, and mate-finding. The physical features and substrates enable custom conditioning of harborages at the species and conspecific level. A secure and established harborage can facilitate the efficiency of short-distance sex pheromones and improve mate-finding in small populations. Conspecifics can respond to aggregation pheromones and strengthen the "harborage scent" that secures the space from others. Foraging efficiency to and from feeding sites is increased by pheromones that guide the movement. The network of harborages in large populations provides a secure foundation for persistence, population increase, and dispersal.

Harborage features, such as size, substrate, and surface texture, influence their utility for some species. The behavior of self-crowding in voids (thigmotaxis) can make almost any space suitable (Fig. 5.1). But suitability can be determined by the size of the opening and the texture of the interior surface. A rough surface is preferred by bed bugs, which rest in the harborage for long periods between a blood meal. These

Fig. 5.1. German cockroach female. Public Domain.

© William H. Robinson 2026. *Pests in the Urban Biome* (W.H. Robinson)
DOI: 10.1079/9781800626416.0005

insects do not have pads on the tarsal segments of their legs and depend on the large tarsal claws for a secure hold on a rough surface. The common bed bug has difficulty climbing smooth vertical surfaces without these pads, while the tropical bed bug has a feature on the legs that enables this climbing movement.

Cat Flea

The cat flea (Fig. 5.2) infests domestic dogs and cats worldwide. Mammals in urban environments, such as foxes, coyotes, raccoons, opossums, and skunks, are also hosts. The cat flea is more common than the dog flea, and it occurs on both animals. Recently emerged adult fleas indoors will orient toward humans and attempt to feed. However, feeding on humans is limited because human skin is thicker than that of a dog or cat and difficult for flea mouthparts to penetrate.

Life cycle

A blood meal is essential for female fleas to produce eggs. Adult fleas eliminate 8–10 droplets of blood as feces during each feeding. This blood-feces dries in the hair of the host and falls where the pet rests. Females begin to produce eggs about 2 days after their first blood meal. A female lays about 14 eggs per day, and the lifetime total is about 150. eggs hatch within 4 days. They have a smooth outer shell and easily fall from the host after they are laid. In the first larval stage, mouthparts enable them to feed on the hardened blood feces. The third-stage larva spins a silken cocoon and molts to the pupa

Fig. 5.2. Cat flea adult. Credit: W. Robinson.

stage. Later, it molts to an adult flea inside the cocoon. The pre-emerged adult can remain in the cocoon up to 140 days before emerging. The life cycle is completed in about 14 days (Fig. 5.3).

Harborage

Indoor carpeting provides an ideal harborage for cat flea infestations (Fig. 5.4). The fleas on a dog or cat provide a significant portion of the food requirements for their offspring. The nutritional value of the blood-feces is high and used by the immature stages during their development. The cat flea has a unique form of parent–larva investment. Females deposit eggs that accumulate in the harborage, and the adult fleas deposit a food resource in the resting site. Larvae feed on the dried feces, and the protein content of this dried blood is essential for development. Cat fleas have periods when they produce nonviable eggs. Nonviable eggs accumulate in the harborage along with viable eggs. Successful development of adult fleas is from larvae that feed on a mix of dried blood feces and nonviable eggs. Low (46%) adult emergence results from larvae feeding only on dried blood; larvae feeding only on nonviable eggs have even lower (13%) adult emergence.

The unique aspect of the cat flea harborage is the nutritional investment one generation has in the next. All stages are involved: flea eggs have a smooth shell and easily fall from the host after they are deposited, dried blood feces falls from the host, and larvae have the mouthparts to feed on the hardened blood fragments and the collapsed unfertilized eggs.

Harborage substrate

The most important factor influencing the success of the cat flea indoors may be the widespread use of pile carpeting in modern households. This substrate is an excellent substitute for the nest habitat of its natural animal hosts and provides larvae with the humidity and temperature conditions needed for development. Carpet pile acts to retain and concentrate the food necessary for the larvae, and it provides a substrate for the pupal stage.

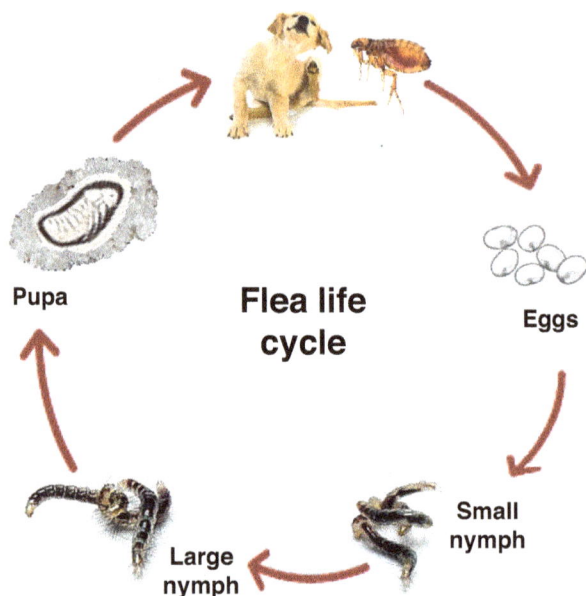

Fig. 5.3. Cat flea life cycle. Credit: W. Robinson.

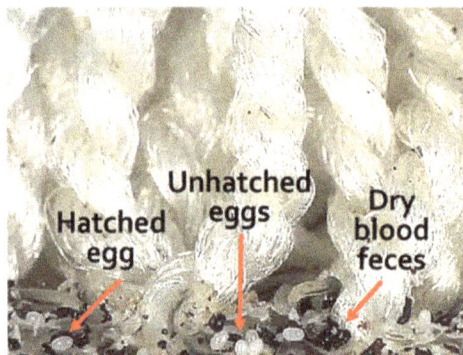

Fig. 5.4. Carpet cross-section showing flea eggs and adult blood feces. Credit: W. Robinson.

Distribution

Flea eggs fall from the host soon after they are laid or when the host arises after sleeping. Flea larvae are usually concentrated in areas where the eggs are, and are absent from other areas. The presence or absence of flea larvae in an area can depend on pedestrian activity, proximity to a pet sleeping area, and abundance of food. The location of flea larvae in carpeting in a bedroom of a house with one companion dog illustrates the influence of the above factors on their distribution (Fig. 5.5).

The dog frequently slept on the bed at night. In the morning it would jump from the bed to the floor then exit the room. The force of impact onto the floor dislodges flea eggs deposited during the night. This daily routine of sleeping on the bed and jumping to the floor results in flea eggs and dried blood feces accumulating in the carpet at the end of the bed close to the door.

Distribution and location of the flea larvae in the carpet is similar to that of the eggs, which indicates that larvae do not move far from the site of hatching. Host blood in the form of adult flea feces is essential to larval cat flea survival and development. The supply of food in the carpet where first-stage larvae hatch is assured because both eggs and flea feces drop from the pet at the same time and collect at the same location. The distribution of third-stage larvae was similar to that of the first- and second-stage larvae.

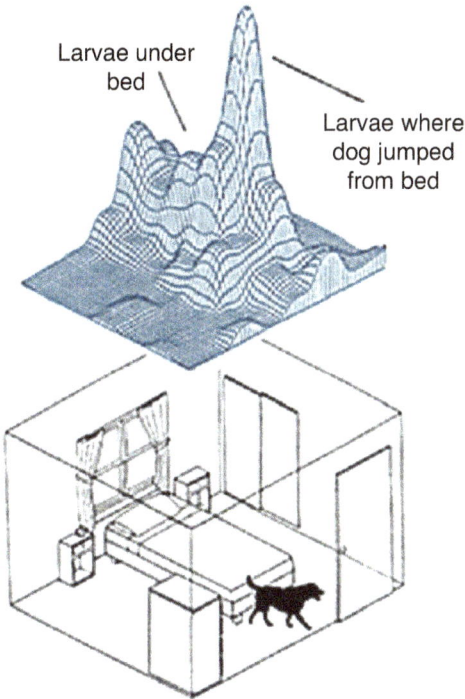

Fig. 5.5. Flea distribution in house bedroom.
Credit: W. Robinson.

Fig. 5.6. Mid leg and hind leg of adult flea. Credit:
W. Robinson.

Leg setae

Adult flea legs are adapted to moving through
the dense hair coat or feathers of animals
(Fig. 5.6). The legs have large and small setae
that are directed backward and located on the
posterior surface of the leg segments. This
position facilitates moving forward because
the setae do not engage hairs. But a small
movement backward by the flea engages the
setae with hairs. This feature turns the leg and
body setae into effective hold-fast structures
that hold the flea to the host.

Bed Bug

Bed bugs are blood-sucking insects that occur
around the world. The common bed bug, *Cimex*

lectularius, has worldwide distribution, while the
tropical bed bug, *C. hemipterus*, occurs primarily
in tropical and subtropical regions (Fig. 5.7).
There are no natural populations of these
species; they are confined to indoor habitats. Bed
bugs do not spread disease to humans.

Life cycle

Adults and nymphs locate a host by detecting
body heat and CO_2. Females require a blood meal
to lay eggs; they produce up to 20 eggs from a
single meal. They can lay several eggs per day
in secluded places; eggs hatch in about a week.
Nymphs molt five times before becoming an
adult, and a blood meal is needed between each
molt. Nymphs and adults can survive for several
months without feeding. After feeding, they are
vulnerable to being crushed by movements of
the host and leave the feeding site immediately.
Adults and nymphs remain in harborages when
not foraging. Infested harborages are close to
a potential host; adults and nymphs leave and
return to the same harborage (Fig. 5.8).

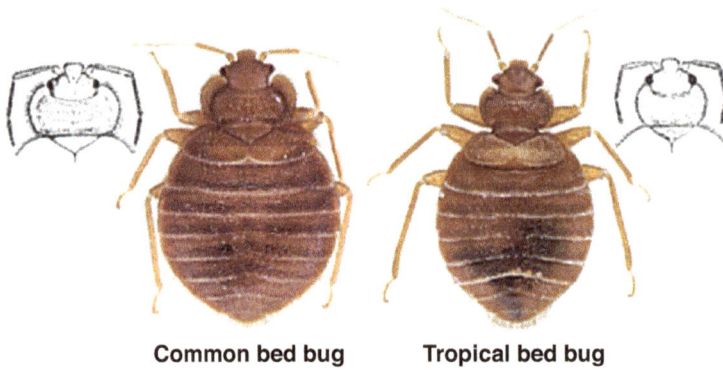

Fig. 5.7. Common bed bug and tropical bed bug. Credit: W. Robinson.

Common bed bug   **Tropical bed bug**

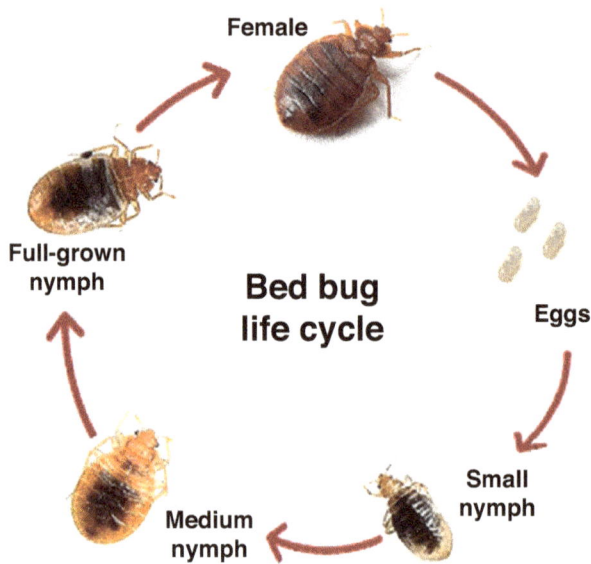

Fig. 5.8. Bed bug life cycle. Credit: Pixabay/CC0 Public Domain.

Harborages and infestations

Bed bug harborages usually contain adults, eggs, nymphs, feces, and an accumulation of exuviae from developing nymphs (Fig. 5.9). Bed bugs return to harborages after feeding and remain until the next foraging event. Attraction to volatile pheromones directs their movement toward existing harborages where related individuals are developing. A volatile aggregation pheromone in the harborage attracts adults and nymphs from a distance. A compound present in feces and the shed nymphal skins in the harborage functions to retain bed bugs in the harborage.

The harborage is key to bed bug feeding, reproduction, and survival of populations in the household habitat. They prefer to occupy narrow spaces and to rest on rough surfaces. The initial narrow void selected as a harborage is gradually conditioned with pheromones to transform it into a suitable harborage for adults, nymphs, and eggs. Narrow spaces may have limited air currents, and crowding of individuals in the small space increases the humidity. This reduces water loss during long periods of inactivity and improves the habitat for nymphal development. Feeding on the host takes only a short time, and bed bugs leave the host immediately after

Fig. 5.9. Bed bug female and cast skins of nymphs in harborage. Credit: W. Robinson.

Fig. 5.10. Bed bug hind leg. Credit: W. Robinson.

the event. After feeding, adults and nymphs switch from host-seeking to harborage-seeking behavior. This is guided by a volatile aggregation pheromone in the harborage. Bed bugs may follow chemical trails on surfaces to and from their hosts.

The origin of individual bed bug infestations may be a single gravid female or a small group of conspecifics, perhaps originating from the same harborage. There seems to be limited exchange of adults or nymphs between infestations in the same structure. Isolated populations increase through cycles of inbreeding, which results in increased fitness for the urban environment.

Foraging

The central-place foraging strategy of bed bugs includes detecting chemicals that guide them to their food and detecting chemicals that guide them back to their harborage. They spend limited time obtaining a blood meal and most of their time inside the harborage between feeding events. A compound present in the feces and on the shed (after molting) skins of nymphs in the

harborage acts to retain bed bugs in their harborage. Bed bugs aggregate in many locations within the host's sleeping room, but usually 10–15 m from where the host sleeps. Although human skin odors attract bed bugs, human skin compounds also prevent them from remaining on hosts after feeding. About 8 days after a blood meal, a female has digested most of the blood and laid most of the eggs developed from that meal. At that time, females leave the harborage and seek another host.

Leg setae and tarsal pads

A strong preference for rough surfaces may be linked to the absence of large tarsal pads on the legs and an inability to grasp smooth surfaces (Fig. 5.10). The terminal leg segments (tibia) of bed bugs are tubular and narrow; there are no tarsal pads on the underside and there is no pad between the large tarsal claws. But there are a few large setae along the outer edge. The tarsi segments fit together to form smooth contours that may not impede movement through substrates or on the host skin surface. The absence of tarsal pads limits walking and climbing on smooth surfaces to which tarsal pads would adhere. The large claws on the end of each leg provide a grip on rough indoor surfaces and on similar harborage surfaces.

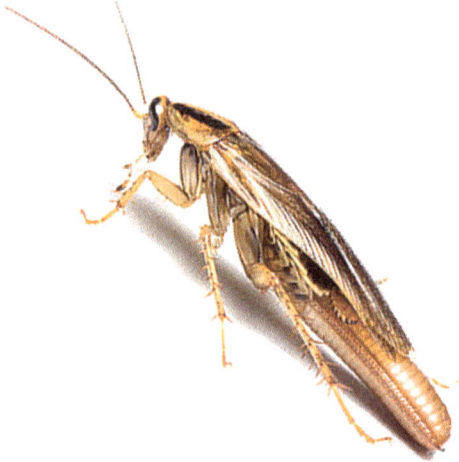

Fig. 5.11. German cockroach female. Public Domain.

German Cockroach

The German cockroach (*Blatella germanica*) is the most recognized indoor pest worldwide (Fig. 5.11). Small body size, foraging mobility, omnivorous feeding, and reproductive habits provide for its success. The ability of populations to develop physiological or behavioral resistance to insecticides often limits control measures and increases their pest status. This cockroach lives only indoors and does not exist in natural populations.

Life cycle

The gravid female carries the eggcase until about 24 hour before it hatches and the nymphs emerge. Females remain in a harborage while carrying the eggcase but periodically leave to feed and drink prior to depositing the eggcase. The first-stage nymphs usually remain in the harborage and feed on adult fecal deposits on the interior surface. Females need to mate only once to produce all six or seven eggcases. The number of nymphs per eggcase ranges from 34 to 47. The life cycle takes about 100 days. Breeding is continuous, with overlapping generations present at any one time in an infested location (Fig. 5.12).

Harborage

The harborage is key to the reproduction and survival of the German cockroach (Fig. 5.13). This synanthropic species selects voids in the habitat that meet the physical size and substrate preference for adults and nymphs that will occupy the space. The original empty space is gradually transformed into a long-term functional living and breeding space. Pheromones and other volatile chemicals produced and applied in the harborage by adults and nymphs are linked to the functions of the space. These include aggregation of nymphs and adults, attraction of conspecifics to the space, and repellency of others. The harborage is initially conditioned and then maintained as a central place or anchor for a large population of conspecifics. It is key to the success of this species in the urban biome.

New harborages are established by males and mid-size nymphs, as these individuals forage away from their initial harborage and may encounter eligible sites. Narrow entry (3.2–6.2 mm) voids near food and water are preferred because these are essential resources. The narrow opening, large internal size, and a rough surface are important features. Inside, the age and gender groups segregate by body size to use the height and depth of the space. Small nymphs remain in the narrow sections of the harborage, and adults are typically together where it may be larger. The internal environment supports the young nymphs. They have a high respiratory rate and limited movement and depend on the humidity created by a dense aggregation to reduce body water loss. Adult feces available on the substrate provides food for young nymphs and reduces the need to forage outside the harborage.

Conditioning

The German cockroach harborage is a chemically conditioned space specifically for mating, egg laying, and developing immatures. The combination of selecting and transforming a space into a multifunctional harborage, rather than taking it as it is, has ensured the

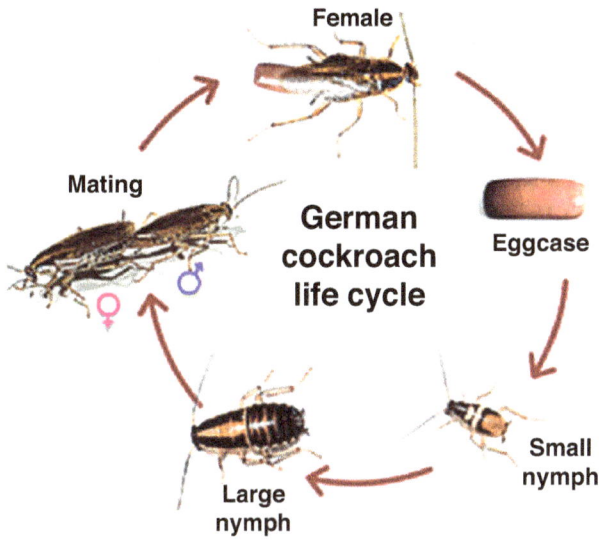

Fig. 5.12. German cockroach life cycle. Credit: W. Robinson.

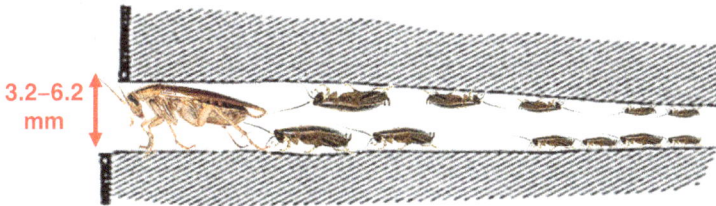

Fig. 5.13. German cockroach harborage graphic. Credit: W. Robinson.

German cockroach as a successful synanthrope. Pheromones increase the efficiency of forging, and reusing food eaten by adults by feeding early nymphs on their feces further improves efficiency. These traits maintain the stability and habitat dominance of the German cockroach. Pheromones that control harborage activities include contact chemicals and air-borne volatiles which function as sex pheromones, attractants, repellants, dispersants, alarm pheromones, and trail pheromones, and facilitate kin recognition. A chemical in the adult feces stimulates aggregation behavior. The continued presence of females and males in the harborage ensures mating with conspecifics. This harborage has unique value when compared to other cockroach species. American and oriental cockroaches do little to condition their harborage, which is basically a temporary shelter.

Foraging

German cockroaches forage individually, but conspecifics often converge at the same food sites (Fig. 5.14). A trail pheromone may direct movement from the harborage to a food source. Adults and nymphs leave traces of saliva and fecal deposits at the feeding sites. These residues may orient conspecifics to the site and reinforce the harborage scent by their feces when they return there. The harborage remains at the center of foraging activity by orienting the return and benefiting from the feces deposited between foraging events.

Dispersal

Pheromones in the saliva of the German cockroach act to disperse adults and nymphs. This

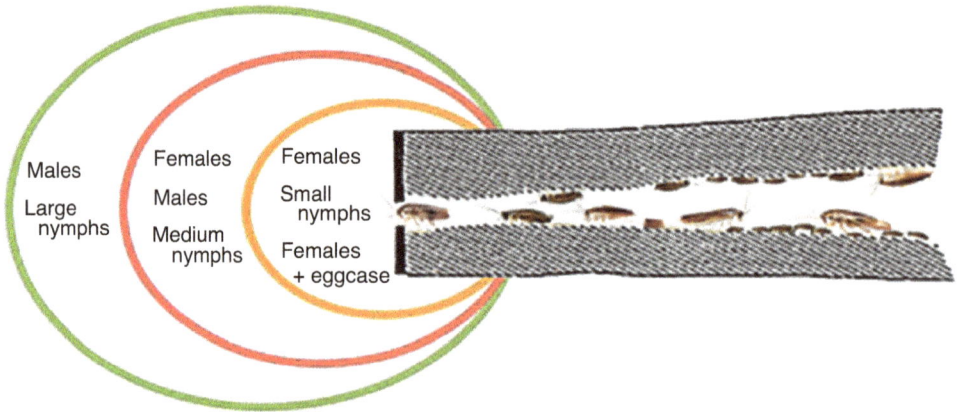

Fig. 5.14. German cockroach foraging range. Credit: W. Robinson.

Fig. 5.15. German cockroach hind leg tarsal segments. Credit: W. Robinson.

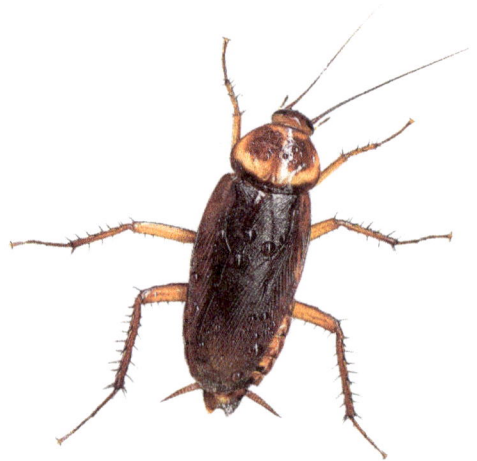

Fig. 5.16. American cockroach adult male. Public Domain.

pheromone counteracts fecal attractants and is concentrated in the saliva of crowded, gravid females. It is thought to function as a space regulator within aggregations. Deleterious effects from crowding begin to occur in *B. germanica* when they exceed a level of 1.2 individuals per cm^2 in a harborage. The aggregation of individuals strengthens the "scent signature" of the harborage and the conspecific group. The scent is linked to the gut bacteria acquired by conspecifics feeding on the same foods in an area surrounding the harborage. The scent is repellent to non-specifics and limits reproduction outside the group.

Legs and tarsal pads

B. germanica has pads on the tarsi of each leg (Fig. 5.15). The pads are large and surrounded by spines. The large pad between the claws provides traction for vertical or smooth surfaces. Tarsal pads on females are 1.5 times larger than the male's.

American Cockroach

This large (~4 cm) cockroach (*Periplaneta americana*) lives primarily outdoors as well as indoors in habitats from urban landfills, steam tunnels, and building basements to underground sewer systems of the major cities of the world (Fig. 5.16). Natural habitats include moist areas in wet leaf litter and undisturbed areas with

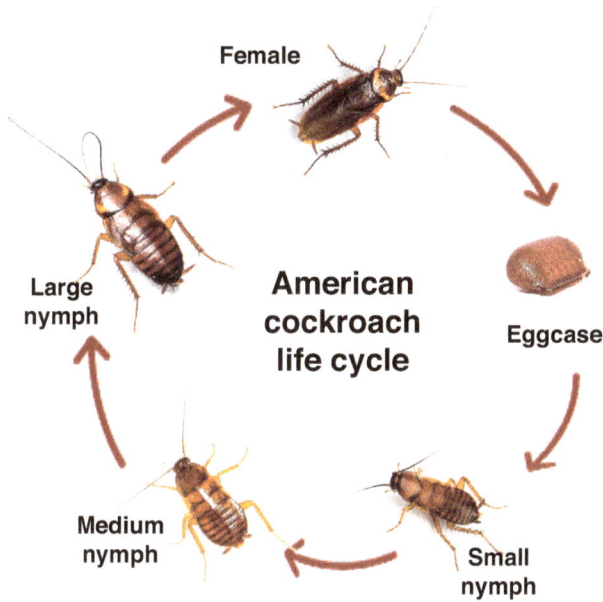

Fig. 5.17. American cockroach life cycle. Credit: W. Robinson.

dense vegetation. Adults may occur indoors while foraging from outdoor sites or in building basement sites with connections to a sanitary sewer system.

Life cycle

Eggcases contain 16 eggs, and females carry them for about 24 hour; 10–15 eggcases are produced within 10 months. When deposited, eggcases may be secured to the substrate by an oral secretion and sometimes covered by debris nearby. Fresh mating is not necessary for each eggcase; females can produce their full potential from a single mating. Nymph development is through 7–13 instars for about 15 months (Fig. 5.17). Preferred temperature for adults and nymphs is about 28°C. Adults fly when the temperature is above 21°C. They will fly to outdoor lights at night.

The concept of harborage is expanded with the *P. americana* populations in urban environments. For this species, the municipal sewer systems and harborage can be considered as one. There the environmental conditions are relatively uniform throughout, and the American cockroach population can thrive. While the German cockroach prepares and conditions a distinct space, the American cockroach seems to adjust to the sewer system conditions without making changes. There seems to be no separation into sites for food, foraging, and deposition of eggcases. Leaving and returning from the system is from street openings (covers) and through pipes that connect to building basements. The female uses a sex attractant pheromone that fits well in the humid airspace and limited air currents in sewers.

The American cockroach is well adapted to indoors and outdoors, but its large size limits the amount of secluded harborage indoors. It can escape danger with speed and strength. Adults can run 40–50 body lengths per second, and on any surface. At high speeds, adults run on four legs or raise their body and run forward on two legs. They can squeeze through gaps of 4–6 mm while still maintaining relatively high running speed. Adults can fly as fast as they run and easily negotiate short distances indoors. The mouthparts have crusher-cutter mandibles with a bite force 50 times stronger than the adult body weight.

Legs and tarsal pads

The pads on the tarsal segments are large and secrete a liquid that increases contact and grip on substrates (Fig. 5.18). The pad between the large claws on the last segment engages smooth surfaces when climbing or crawling indoors or on vegetation outdoors. The mid and hind legs have long setae or spines that enable quick movement on the uneven surfaces of building basement walls and the inside walls of sewer pipes and steam tunnels. This cockroach often occurs around the perimeter of buildings, and the long setae on the legs support this distribution.

Habitat

This species is restricted to warm (21–33°C) and humid habitats. The body covering, the cuticle, is highly permeable and makes this cockroach susceptible to water loss. The permeability of the cuticle limits its distribution in humid locations. Optimum relative humidity for adults and nymphs is above 70%, which is sufficient to prevent excess water loss, but not likely to occur indoors. The American cockroach has benefited from urbanization and improved housing conditions and modern urban sewer systems (Fig. 5.19). Central heating in city buildings and residential units changed the ambient temperature in the sewer system. Water entering sewers from heated buildings is around 20°C and remains at a uniform temperature as it travels to the treatment facilities. This feature maintains humid conditions in underground sewer pipes and likely secures this cockroach access to urban sewer systems around the world.

The continuous production of eggcases and the short time before deposit frees females for movement away from the harborage. Most movements away from sewers are within the range of 20 m. In sewer system manholes, resources and conditions are relatively stable year-round, and dispersal to other habitats is by sewer connections to buildings or street openings (Fig. 5.20). Adults and nymphs travel over a wide area that includes other manholes or the interconnected system of pipes. Individuals disperse by walking and flying to new sites above ground. In sewer populations, more females move to new sites compared with males. An aggregation pheromone in the feces promotes concentrations of adults and nymphs in harborages.

Considering the sewer habitat is regularly occupied by American cockroaches, a significantly important attribute is resistance to disease. This species has more DNA than almost any other insect, and with 3.3 billion DNA bases, it is comparable in size to the human genome. A part of that genome is a set of Toll genes. These provide immunity and the ability to synthesize and secrete antimicrobial molecules. Sewer pipes contain both biofilm and sediment with microbes involved in the breakdown of organic compounds. The strong immune system of American cockroaches may help them survive in unsanitary sewer conditions.

Food resource

Foraging activity in urban habitats continues throughout the year, but there are periods of limited activity. In habitats with seasonal changes in temperature and humidity there is a foraging pattern that correlates with these changes. Foraging generally decreases from December to February and increases in March and April, then remains unchanged for the next 6 months. As a result of this cycle, the body weight of males and females is highest in December, which is at the end of the 6-month

Fig. 5.18. American cockroach hind leg tarsal segments. Credit: W. Robinson.

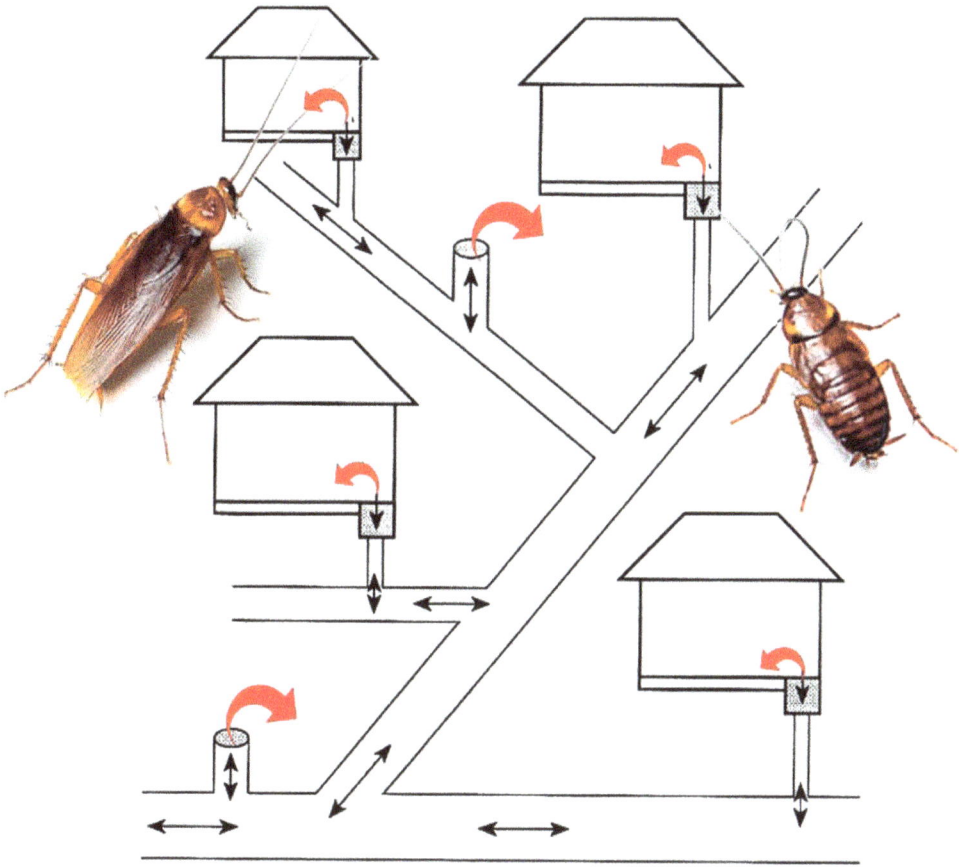

Fig. 5.19. American cockroach in urban sewer system. Credit: W. Robinson.

foraging period, and lowest after the 3-month period of limited foraging.

sites, such as behind picture frames or in the corners of interior doors.

Brownbanded Cockroach

This small (~14 mm) cockroach (*Supella longipalpa*) (Fig. 5.21) is similar in size and habits to the German cockroach. In the urban biome, it can live outdoors and indoors. In warm climates, it is abundant year-round; indoor populations are successful in a range of climates. In buildings and houses, adults and nymphs are not limited to kitchens and bathrooms but often infest other rooms. They prefer resting locations high on the walls of heated rooms and deposit eggcases in elevated

Life cycle

Eggcases contain 14–18 eggs, and hatching occurs in about 37 days at 30°C. The number of eggs per eggcase remains nearly unchanged during the life of the female, and 10–20 eggcases are produced. The eggcase is deposited about 24 hour after it develops and the female glues it to a substrate. eggcases are often placed at the same location by many females, and there may be clusters behind picture frames. There are 6–8 nymph stages for males and females. Adult lifespan is about

Fig. 5.20. American cockroaches in sewer manhole. Credit: Encino-Tarzana, CA.

115 days. Adults mate several days after maturation, and the first eggcase appears in about 10 days. Indoor population size has limited fluctuation during a 12 month period. There are increases in numbers of adults and nymphs during warm months. Dispersal is primarily by adults, but large nymphs have limited movement. Adult males can fly both indoors and outdoors; they commonly will hop short distances when indoors.

Legs and tarsal pads

The adults and nymphs have small tarsal pads on their legs, but the pad between the terminal claws is large (Fig. 5.22). While the German cockroach has tarsal pads on all legs but primarily uses the tarsal claws to climb vertical surfaces, the claws and small tarsal pads of the brownbanded cockroach are sufficient to provide the ability to climb vertical surfaces, such as walls. eggcases are often placed in secluded locations that allow access by the small pads. The small tarsal pads on the brownbanded cockroach may limit exposure to liquid and dust insecticides applied to surfaces for control.

Harborage

The brownbanded cockroach does not establish or use a secluded or enclosed harborage

Fig. 5.21. Brownbanded cockroach adult male (top) and female (bottom). Credit: USDA.

Fig. 5.22. Brownbanded cockroach hind leg. Credit: W. Robinson.

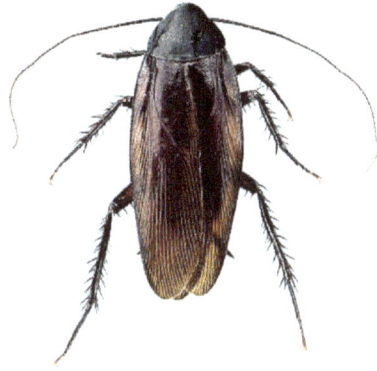

Fig. 5.23. Smokybrown cockroach adult male. Public Domain.

space to rest when not foraging. Sites used for aggregation are structural features of the habitat, such as corners or doors of cabinets. The surfaces of these sites have fecal smears, and aggregation of adults may be linked to the residue of a female sex pheromone (supellapyrone). Females deposit eggcases away from these aggregation sites, and the large number of eggcases deposited by individual females may be linked to a volatile attractant.

Smokybrown Cockroach

The smokybrown cockroach (*Validiblatta fuliginosa*) (Fig. 5.23) occurs in urban and natural sites primarily in warm climate regions. It is not cold-tolerant and often moves indoors to avoid seasonal cold temperatures. It lives and forages along the perimeter of buildings, especially places where there is vegetation shade and moist groundcover. The pest status of this cockroach is based on its occasional entry into houses and overwintering in houses, such as in attics and wall voids. Global distribution includes southern regions of the USA, Asian countries, Australia, South America, and Europe.

Life cycle

Females emit a sex pheromone called periplanone that results in long-range attraction and courtship behavior in males. eggcases contain 20–28 eggs and are formed soon after mating and deposited about 24 hours after formation. Females often prepare an oviposition

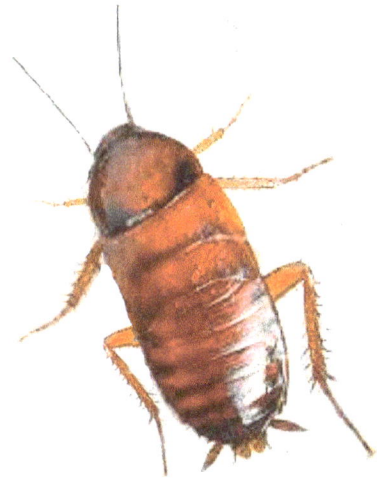

Fig. 5.24. Smokybrown cockroach nymph. Credit: W. Robinson.

site by chewing a hole in a soft substrate; the eggcase is then deposited and partially covered with debris. Females produce about 20 eggcases in their life. There are 9–12 nymph development stages (Fig. 5.24) for males and females. Adults live about 24 months.

Legs, setae, and tarsal pads

Large setae on the femur of the legs may provide some grip for movement in the ground cover of outdoor habitats (Fig. 5.25). The long first tarsal segment may elevate the body

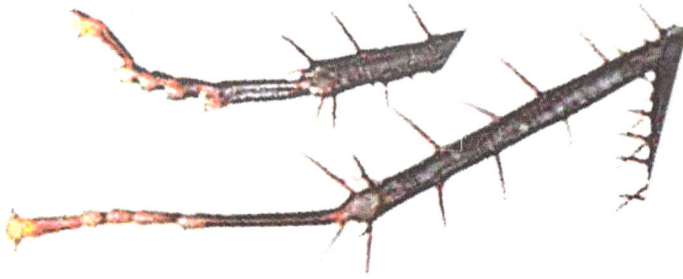

Fig. 5.25. Smokybrown cockroach midleg and hind leg. Credit: W. Robinson.

above ground vegetation. The tarsal pads are reduced, especially on the hind legs. There may be limited need for adults and nymphs to climb smooth surfaces, and small pads may reduce water loss.

Habitat

Outdoor habitats include tree holes, under the bark of trees, or leaf litter. Much of their habitat selection and survival is linked to maintaining their body water balance. The cuticle of the smokybrown cockroach is very permeable to water. It has the highest rate of cuticular water loss of all the cockroach pest species. Because of this vulnerability, the smokybrown cockroach requires moist habitats with available water. In the urban environment, it occurs primarily outdoors among landscaping vegetation. Infestations may be inside structures, but usually there is exchange with outdoor populations.

Populations are successful in habitats with high (>75%) relative humidity and temperature. There is a shift to nymphs in the population when adults die in fall (Fig. 5.26). Overwintering temperatures and low humidity limit movement and feeding. Development of adults in early spring coincides with increased temperature and humidity.

Oriental Cockroach

This cockroach is distinguished by a shiny, black body and its slow-moving gait (Fig. 5.27). The wings of the male cover only part of abdomen, female wings extend only slightly past the thorax; neither sex is capable of flying. Tarsi of adult females and nymphs have a small pad between the claws and last tarsal segment. Despite the common and scientific name, *Blatta orientalis*, this species is not native to the Orient.

Life cycle

Eggcases are produced at intervals of 1–2 weeks; females produce 6 to 8 in total. Formation of the eggcase is completed in 24 hour and it is carried usually for 1–2 days before being deposited. When deposited, it is attached to the substrate or placed in a protected location. Nymphs develop through 7–10 instars at 30°C. Adults appear in May or June and die in July or August. Survival without food or water at 27°C is about 11 days for males and 13 days for females; survival without water is 20 days for males and 32 days for females.

Habitat

Natural habitats for the oriental cockroach include leaf litter and debris in areas with warm summer temperatures and moderate winter temperatures. The success of this cockroach in the urban biome is linked to the deposition, development, and hatch of the eggcase (Fig. 5.28). The success of the eggcase is limited to temperatures of about 20°C, and relative humidity below 75%. Females do not produce eggcases in environments where the temperature is too low (15°C

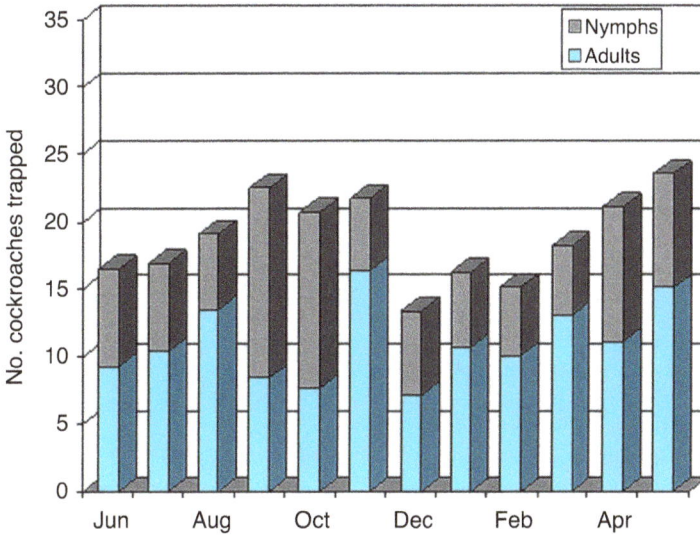

Fig. 5.26. Smokybrown cockroach seasonal abundance in eastern USA. Credit: W. Robinson.

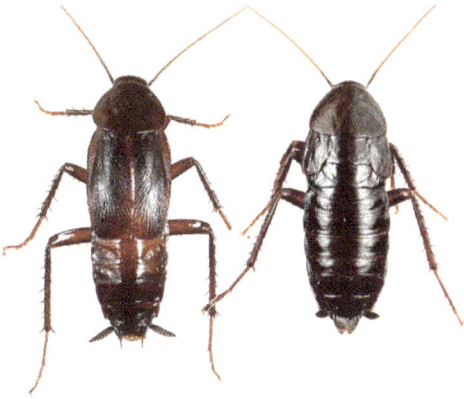

Fig. 5.27. Oriental cockroach male (left) and female (right). Public Domain.

Fig. 5.28. Oriental cockroach female with eggcase. Credit: W. Robinson.

and below) for the eggs to develop and hatch. The number of nymphs emerging from eggcases decreases as relative humidity decreases. This may be linked to the desiccation and hardening of the eggcase surface preventing nymphs from forcing the eggcase open when they are ready to leave it.

This cockroach usually occurs around the perimeter of buildings in the urban core zone. There is not an established harborage either inside or outside. It tolerates a range of temperatures and humidity and can adjust to outdoor winter conditions. Its limited dispersal capabilities may reduce the habitats the adults and large nymphs explore and occupy.

Legs and tarsal pads

The adults and nymphs do not have large tarsal pads on their legs, and the pad between the terminal claws is almost absent (Fig. 5.29). The claws

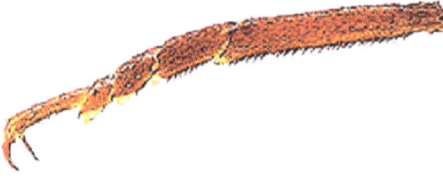

Fig. 5.29. Oriental cockroach hind leg. Credit: W. Robinson.

Fig. 5.30. Common yellowjacket wasp. Credit: Pixabay/CC0 Public Domain.

provide the ability to climb some vertical surfaces, such as rough-surface walls. The absence of a large pad between the claws nearly prevents movement on smooth vertical surfaces, such as those found in bathroom sinks and tubs. Adults and nymphs may accidentally enter and become trapped in these sites, which gives the impression they entered from the drain. The name "water-bug" is based on this misunderstanding of how these cockroaches got in the sink or tub.

Common Yellowjacket and German Yellowjacket

The term "yellowjacket" refers to different species, including the common yellowjacket (Fig. 5.30) and the German yellowjacket. These yellow and black wasps feed on other insects, such as caterpillars, flies, and bees, but they will also feed on fruits and carrion. Their foraging area is about 2 km from the nest site. These two species are abundant in urban environments, including green spaces and habitats around buildings. The most suitable temperature for foraging and daily activity is above 20°C. Continuing global climate change and the warm and cool season temperatures provided by the urban heat island effect will benefit populations of these species.

Colony cycle

The colony starts in spring when the founding queen emerges from an overwintering site. She collects and chews wood fiber and mixes it with saliva to make the paper-like covering of the nest. Then she lays eggs in the initial small nest and rears the first brood of workers. If the spring weather is cloudy and rainy, there will be few insects for her to collect as food for larvae and the colony may fail. In late summer, the colony peaks and the development of workers ceases, and queen cells for the next generation are built. The mature colony consists of 2000–4000 workers. At the seasonal end of the colony cycle, males and queens mate, and the queens select a protective overwintering site to ensure survival until spring, when the cycle repeats.

The common yellowjacket builds nests in soil cavities or an old mouse burrow underground (Fig. 5.31). The German yellowjacket is a void-nesting species that builds in walls, attics, and under eaves. Bald-faced hornets, which are yellowjackets, build aerial nests. The thick envelope of these nests prevents heat from escaping; the optimal temperature of the nest is about 32°C. In suburban areas, skunks will dig up and feed on the larval cells of in-ground nests.

Harborage

The yellowjacket nest is a nightly aggregation site for workers, a larval rearing facility, a shelter from adverse environmental conditions, and a protected space for conspecifics. The primary function of this seasonal harborage is to produce the next generation. The yellowjacket nest serves as an anchor for a large population of conspecifics that forage in a space surrounding the nest. These wasps use pheromones to identify colony members and to deter others from entering the nest.

Fig. 5.31. In-ground and above-ground yellowjacket nests. Credit: Pixabay/CC0 Public Domain.

Invasive

The common yellowjacket and German yellow-jacket are successful invasive species and have spread to several regions of the world over the past century. Despite their similar biology, these two species have developed foraging strategies and food preferences to minimize competition when invading countries.

European Paper Wasp and Northern Paper Wasp

The European paper wasp (Fig. 5.32) occurs throughout most of Europe, across north Africa, and eastward across Asia into China. It was introduced into North America in 1978 and is now distributed across the continent. It has similar coloring to some yellowjackets and can be confused with local yellowjacket species. The Northern paper wasp (Fig. 5.33)

is distributed from southern Canada to the southern USA, and into Central America.

Colony cycle

Fertilized queens overwinter in protected locations. They emerge in spring and create new nests, which are a small cluster of open cells attached to a structure by a short stalk. The nest material is chewed wood and not paper. The queen deposits one egg in each nest cell; larvae hatch in a few days and are fed by the queen. After the first brood of female workers emerge, the size of the nest and number of workers continues to increase until mid- to late summer. At the end of the colony cycle, adult male paper wasps will mate with new queens of other nests.

Paper wasps will vigorously defend their nests, especially late in the summer and fall. There are individual wasps at the opening of

Fig. 5.32. European paper wasp. Credit: Pixabay/CC0 Public Domain.

Fig. 5.34. Paper wasp nest with first-brood workers. Credit: Pixabay/CC0 Public Domain.

Fig. 5.33. Northern paper wasp. Credit: Pixabay/CC0 Public Domain.

species. When conditions are favorable, the first brood of workers begin foraging early and the colony develops quickly (Fig. 5.34). Nest location may be the most important advantage for European paper wasps. They prefer to build nests in enclosed sites, which protects them from bird predation and exposure to rain. Nesting sites along the roof line and around windows are selected when there are limited natural areas.

the nest, and they are most likely to sting when the nest is approached. They aren't aggressive when they are away from their nest.

Foraging

Dispersal

Displacement of native paper wasp species by European paper wasps has been linked to nest building and foraging traits. European paper wasps generally build nests and establish colonies earlier in the season than other paper wasp

On most foraging trips, the workers gather food within a few hundred meters of the nest. Paper wasps are generalist prey foragers but may act as a specialist due to the habit of returning to a specific location. Despite the weather conditions, paper wasps forage daily to maintain developing larvae with fresh food, as larvae fed poor food will produce small workers or die.

Additional Reading

Bao, N. and Robinson, W.H. (2008) Metabolic reserves in *Periplaneta americana* (Dictyoptera: Blattida). In: *Proceedings of the Sixth International Conference on Urban Pests*, ICUP, Budapest, Hungary, pp. 145–152. Available at: https://icup.org.uk/media/5wbhv0cq/icup869.pdf (accessed 29 September 2025).

Bubová, T., Balvín, O., Sasínková, M., Martinů, J., Nazarizadeh, M. *et al.* (2022) Tropical bed bug, *Cimex hemipterus*, established in Central Europe. In: *Proceedings of the Tenth International Conference on Urban Pests*, ICUP, Barcelona, Spain, pp. 315–319. Available at: https://icup.org.uk/media/fwfdqdjz /bubova-173-bed-bug-pp-315-319.pdf (accessed 29 September 2005).

Ferraguti, M., Martínez-De La Puente, J., Brugueras, S., Millet, J.-P., Mercuriali, L. *et al.* (2022) Surveillance and control of mosquitoes in sewers from an urban mediterranean area. In: *Proceedings of the Tenth International Conference on Urban Pests*, ICUP, Barcelona, Spain, pp. 81–85. Available at: https:// icup.org.uk/media/fmsixk3d/14-martina-149-f-pp-81-85.pdf (accessed 29 September 2005).

Hsu, M.-H. and Wu, W.-J. (2002) Effects of nonviable egg consumption on larval cat flea (Siphonaptera: Pulicidae) development. In: Robinson, W. (ed.) *Proceedings of the Fourth International Conference on Urban Pests*, ICUP, Charleston, USA, p. 269. Available at: https://icup.org.uk/media/2luacvjb/ icup279.pdf (accessed 29 September 2025).

Le Patourel, G.N.J. (1993) Environmental aspects of the survival and reproduction of oriental cockroaches (*Blatta orientalis* L.). In: Wildey, K.B. and Robinson, W.H. (eds) *Proceedings of the First International Conference on Urban Pests*, ICUP, Cambridge, UK, pp. 31–34. Available at: https://icup.org.uk/ media/cwuezez1/icup606.pdf (accessed 29 September 2025).

Metzger, M.E. and Rust, M.K. (1999) Studies exploring the overwintering mechanisms of the cat flea, *Ctenocephalides felis* (Bouché) (Siphonaptera: Pulicidae). In: *Proceedings of the Third International Conference on Urban Pests*, ICUP, Prague, Czech Republic, p. 636. Available at: https://www.icup. org.uk/media/1wpg0i3z/icup509.pdf (accessed 29 September 2025).

Pradera, C., Rodríguez-García, E. and Bengoa Paulis, M. (2025) List of synanthropic cockroaches (Blattodea) in Spain and their current distribution. In: Robinson, W.H (ed.) *Proceedings of the Eleventh International Conference on Urban Pests*, ICUP, Lund, Sweden p. 381.

Rust, M.K., Choe, D.-H., Sutherland, A., Sorense, M., Nobua-Behrmann, B. *et al.* (2022) Controlling yellowjackets in urban recreational areas. In: *Proceedings of the Tenth International Conference on Urban Pests*, ICUP, Barcelona, Spain, pp. 45–48. Available at: https://icup.org.uk/media/humonick/ 3-rust-17-f-pp-45-48.pdf (accessed 29 September 2025).

6 Reservoirs

Persistence of most household and structural insects is based on a network of small infestations or populations and large reservoir populations. Species establish their abundance and pest status by initial infestations and maintain their presence and status by replenishment. Pest abundance may decline with isolation of small infestations and the elimination of reservoir populations. The concept of household pest reservoirs parallels the natural foci concept for diseases. These are ecosystems that contains the pathogen and its vectors (animals, insects, ticks), and a reservoir animal that maintains the pathogen's survival. Both insect infestations and human diseases rely on a long-term and stable network of reservoirs that support a network of infestations. Most control or pest management strategies target infestations, but long-term results are often limited.

Habitats with optimal conditions and resources support long-term reservoir populations. Their size and location may be limited in an urban landscape, and their value based on the dispersal habits of individuals or groups (Fig. 6.1). An ongoing exchange with reservoirs can sustain small infestations, and as these increase, the fitness of individuals increases. Pest infestations are usually challenged by ongoing control or management methods. Untreated refugia in the habitat may hold individuals that enable an infestation to recover. But the normal dispersal of reproductives from a reservoir can provide a recovery route. A viable connection with a reservoir population may be evident in the rate of recovery from a disturbance.

Reservoir habitats can also provide a consistent supply of resources and secure breeding sites. For example, urban landfills provide fresh food for insects, birds, and rodents. The steady arrival of organic refuse, and large amounts of human food waste, makes landfill sites an attractive and

Fig. 6.1. Aquatic reservoir on urban street. Credit: Pixabay/CC0 Public Domain.

DOI: 10.1079/9781800626416.0006

routinely visited habitat. Gulls, crows, and starlings make long trips to integrate the food from landfills into their daily diet. Predator birds, such as hawks and eagles, are attracted to landfills, since these sites have increased potential for hunting small birds and nesting sites. The brown rat is commonly associated with landfills but usually feeds and nests close to the site and may not carry food a long distance from it.

Mosquito breeding occurs in natural or undisturbed areas at the edge of urban and suburban areas. Water retained in urban sites, such as along streets and in abandoned building sites, also provides suitable breeding habitats for *Culex* and *Aedes* species. Breeding in these locations can provide the overwintering queens that disperse and establish nests in urban parks or along the perimeter of suburban houses and buildings.

Reservoir Habitats

At various levels and locations, the urban biome provides reservoir habitats for pest species. From some of these reservoirs, such as landfills, adults can reinfest sites that didn't have overwintering populations or colonize new sites. Natural populations established in fringe areas at urban perimeters have the same role and provide individuals to urban sites and harborages.

Landfills: House fly, green bottle fly, and phorid flies

The temperature at the surface of working landfills is close to the ambient air temperature, but below the surface it is in the range of 25–45°C. This high temperature and decaying organic matter support development of house fly and green bottle fly larvae. They prefer to feed and pupate at high temperatures. Adults fly or are wind-carried to surrounding urban areas or transported on refuse-collection vehicles. Phorid flies, including species of *Megaselia*, are tolerant of the temperature and the condition of food waste below the landfill surface.

Residential carpeting: Dust mites and cat fleas

Dust mites inhabit household fabrics and floor coverings (Fig. 6.2). They feed on the human

Fig. 6.2. Pile carpeting microclimate. Credit: W. Robinson.

skin scales that accumulate on these substrates. Persistence of dust mites in the living space is due in part to the potential for reinfestation from reservoir populations. They are capable of dispersing from reservoir habitats on clothing and other fabrics and can be carried to other indoor locations. Reservoir habitats are indoor sites with 75% relative humidity, a temperature of 25°C, and human and pet skin scales. Carpeting that enables a microclimate for skin scales also provides ideal conditions for cat flea eggs, larvae, and pupae.

Sewerage systems: American cockroach

One of the largest cockroach pests, the American cockroach, relies on the largest reservoir population in the urban biome. Megalopolis sewer systems around the world are the reservoir habitat for this species. This continuous and relatively uniform environment is an underground space that extends for kilometers under city streets and supports large populations of this cockroach. Adults exit the system from the manholes to forage and sometimes occupy building perimeters or travel to building basements through sewer pipes. Urban heat islands provide the nighttime temperatures ideal for dispersal.

Fig. 6.3. Subterranean termites (*Reticulitermes* sp.). Credit: North Carolina Extension Publications, Eastern Subterranean Termites 2025.

Fig. 6.4. Subterranean termite (*Reticulitermes* sp.) workers. Credit: University of California, Agriculture and Natural Resources, Pests in the Urban Landscape 2023. https://ucanr.edu/blog/pests-urban-landscape

Natural populations: Termites

Subterranean and drywood termites occur in natural habitats around the world (Fig. 6.3). Once established, they spread to take advantage of the resources, soil moisture, and environmental conditions in urban environments. Transition from natural food to construction lumber includes using mud tubes to connect the colony to above-ground food sources. Indoor infestations may feed throughout cold months and stimulate early swarming behavior. Distribution in urban environments takes advantage of the urban soil temperature and food resources. The density of termites in urban areas can be as many as 25 colonies per 0.4 ha.

Termites

The economic importance of termites as structural pests has led to public awareness of their habits and pest status around the world. Termites utilize cellulose as a food source and feed almost entirely upon seasoned wood, although some species attack living trees. Formosan termites can infest live trees, including oaks, ash, and cypress. Termites are successful because colonies are large, long-lived, and often difficult to detect until damage to structural wood is extensive.

Termites live in colonies comprising specialized individuals, including male and female

reproductives, workers, nymphs, and soldiers (Fig. 6.4). The concept of large colonies with extremely enlarged queens laying thousands of eggs per day is often the public's perception of termites. While true for some tropical species, it is not true for species that inhabit the urban environment.

The wood termites use for food has usually been attacked by fungi and other microorganisms by the time termites find it. The workers forage randomly but are attracted to wood with high moisture (>20%) content and infested by fungi. Reservoir colonies in the natural environment provide winged reproductives that explore habitats in the urban environment.

Distribution

Average daily temperature and soil moisture are important factors affecting the distribution of termites in any ecosystem (Fig. 6.5). Termites are primarily in tropical and subtropical climatic zones because of the uniformity of temperature and soil moisture in these areas. However, *Reticulitermes* species have extended their range north in urban environments, where temperature and soil moisture are often significantly different than the surrounding natural or agricultural environments. Subterranean termites

Subterranean termite distribution

None to light

Light

Light

Moderate

Heavy

Millimeters water per meter of soil

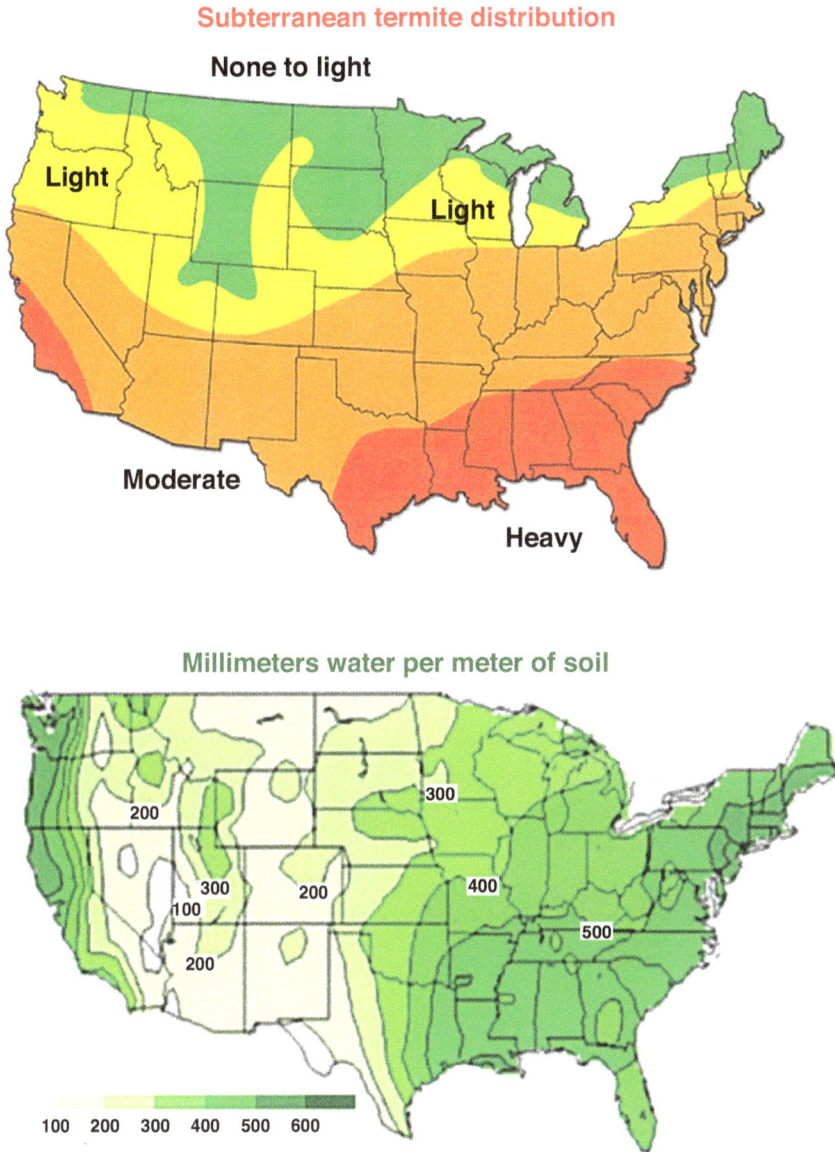

Fig. 6.5. Subterranean termite (*Reticulitermes* spp.) distribution in the USA. Credit: USDA.

are resilient to moisture extremes and can tunnel in soil with a moisture range of 5–30%. Colonies can tolerate low moisture conditions in their foraging environment for extended times, which may allow them to tunnel and find new moisture sources for colony survival.

Modern buildings with concrete slab construction and central heating have contributed to the successful extension of the range of subterranean termites. In temperate regions, termites may be forced to remain deep (1.5 m) in the soil to avoid soil temperatures of 0°C or below during cold months. But there is a limited amount of food and moisture available at these depths. In heated buildings the perimeter walls and foundation slab may provide sufficient heat

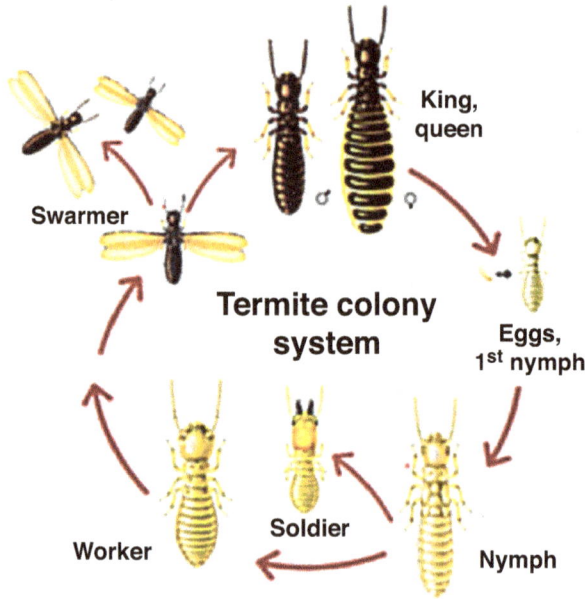

Fig. 6.6. Termite life cycle. Credit: W. Robinson.

to adjacent soil to permit feeding by termites during winter months.

Life cycle

The role of the king and queen is reproduction and initiation of the colony and nest (Fig. 6.6). They mate only once. The queen produces eggs and provides the pheromones that regulate the colony. She can lay more than 3000 eggs per day during her lifetime. However, most eggs are produced by supplementary queens. The nymphs molt several times during their lifetime. It takes more than 4 years for a termite colony to reach the maximum size of 60,000–200,000 workers. The mouthparts of workers are for chewing wood, and workers are responsible for foraging and nest building. Soldiers have large mandibles which enable them to defend the colony against predators. They cannot feed on their own and must be fed partially chewed wood by workers. Workers regularly groom soldiers, and they also groom the queen and supplementary reproductives. When colonies achieve their maximum size, they usually produce swarmers (adults).

Most colonies of subterranean species produce several large swarms in spring (Fig. 6.7).

Fig. 6.7. *Reticulitermes flavipes* termite swarmers. Credit: University of Georgia Field Rpt. C-868 2022.

Rainfall may be the trigger for the start of swarming. Swarms may occur in heated buildings from midwinter into spring. Formosan termites swarm from dusk until midnight from April through July. Swarming is not very effective; only a few of the pairings result in a new colony.

Factors that influence the adaptation of termite species to the urban environment include population density in the surrounding natural

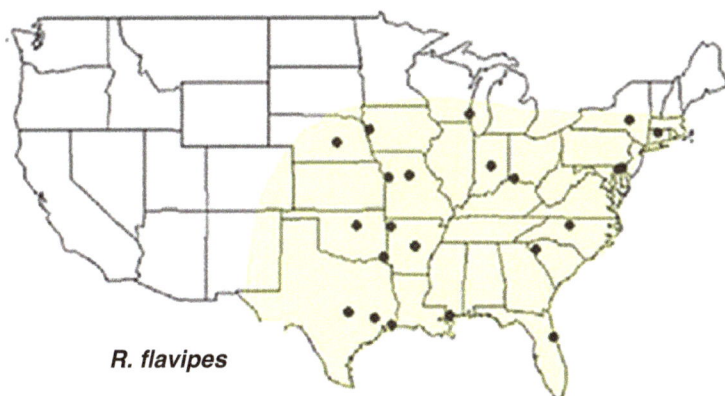

Fig. 6.8. *Reticulitermes flavipes* termite distribution in the USA. Credit: Adapted from Janowiecki *et al.* 2021.

R. flavipes

area, and the existing vegetation and soil type that may provide termites access to a food source. Land that has been in agricultural use before being converted to use for residential or commercial buildings is less likely to have active colonies of subterranean termites than land that was once forested and had natural populations of termites. Preparation of forested land for the construction of buildings includes altering the soil. This often results in removing trees and burying logs, stumps, or other wood debris. This underground cellulose material can be a food source for termites and may contribute to structural infestations. Forested areas adjacent to buildings can be a source for species of drywood termites that normally utilize dead trees and branches as nest sites.

Termites in the urban environment

The termites that attack structural timber are usually categorized as either subterranean or drywood termites. These terms describe the condition of the wood infested or the environment in which the attack occurs. Subterranean termites depend on a moist or humid environment which is provided below the surface of the soil. These termites usually infest wood that has contact with the soil, or wood that has a high moisture content. Drywood termites do not require high levels of moisture and are able to live in dry wood that has no soil contact.

Reticulitermes flavipes

This is the most common subterranean termite species in the urban and suburban environment of north-eastern USA (Fig. 6.8). The northern limit of its range is approximately the line where the minimum temperature does not fall below

Fig. 6.9. Formosan termite workers and soldiers. Credit: USDA

−6°C; southward, its range extends to the Gulf and into Mexico. *R. flavipes* is native to North America, but pest populations are known to have spread to regions in southern Europe.

Coptotermes formosanus

The Formosan subterranean termite is one of the most destructive subterranean termites in the world (Fig. 6.9). Native to mainland China, *C. formosanus* has been distributed by commerce to many areas around the world. Distribution is limited by the inability to tolerate dry periods and high or low temperatures. This species has the potential to develop large and destructive colonies within as short a time as 6 months. The Formosan termite is a major structural pest in China, Tiawan, and Australia. In the USA it is reported in Hawai'i, several south-eastern states, and recently in California.

Mastotermes darwiniensis

This species is the most destructive subterranean termite in Australia. It occurs in both urban and natural environments, and in coastal and inland areas (Fig. 6.10). It attacks a range of structural wood species. *M. darwiniensis* is also known to damage a variety of non-wood household materials, such as plastic cables and leather. It is one of the largest of the subterranean termites; the soldiers are about 13 mm long. Colonies can attain population levels of more than 1 million termites.

Cryptotermes brevis

This species is considered the world's most destructive drywood termite. It is widely distributed throughout tropical and neotropical areas of the world. In has been introduced into the continental USA (Fig. 6.11) and Hawai'i, Australia, South Africa, and Central America. It is a pest in urban and rural areas, and it attacks hardwoods and softwoods used as structural and decorative timbers in houses as well as household furniture.

Ants

Most ants in urban areas have adapted to the soil type and vegetation characteristics of this environment. Few ant species have adapted to the habitat of houses to the extent that they exclusively nest and forage indoors (Fig. 6.12). The reason for this is probably to do with food and nesting habits. Ants do not have a single-source diet, but feed on a variety of organic matter, especially insects and other ants. Nest site selection may require temperature and relative humidity conditions that indoor habitats lack.

Ants are social insects that form small to large colonies (Fig. 6.13). A typical colony contains an egg-laying queen and many workers together in a nest with eggs, larvae, and pupae. Workers provide nest construction and maintenance, forage, and tend the larvae, pupae, and the queen. All workers are sterile females and do not lay eggs. Winged queens and males are present in the nest for only a short period. Queens look like the workers but have larger bodies. Males are about the same size as the workers but have smaller heads and mandibles. In many cases, males look more like small wasps than ants. Workers are the most visible part of the colony, but only about 11% of the workers in a colony are foraging at one time.

The different stages of ant development, such as eggs, larvae, and pupae, occur in the nest. Queens lay the eggs, which hatch in 2 weeks. Workers care for the eggs, feed larvae with food gathered by foraging workers, and care for the pupal cocoons before the adults

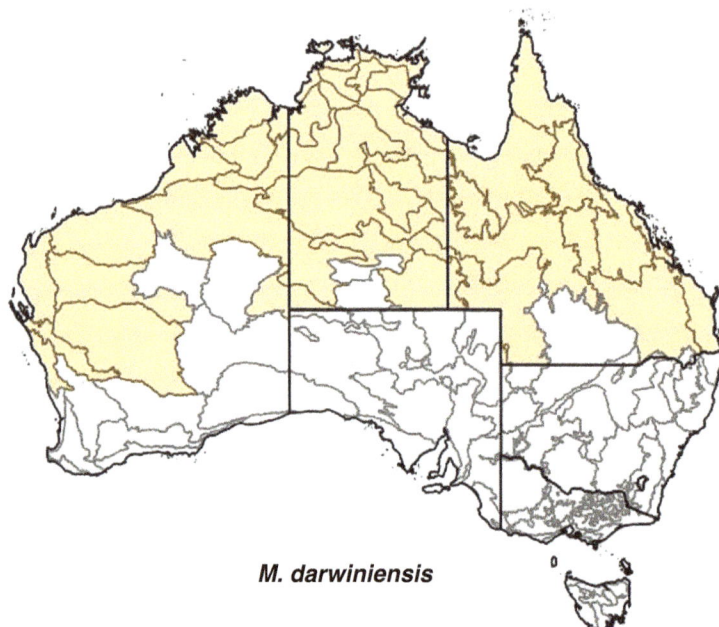

Fig. 6.10. Distribution of *Mastotermes darwiniensis* termite in Australia. Credit: Australian Government, Department of Climate

M. darwiniensis

Drywood termite distribution

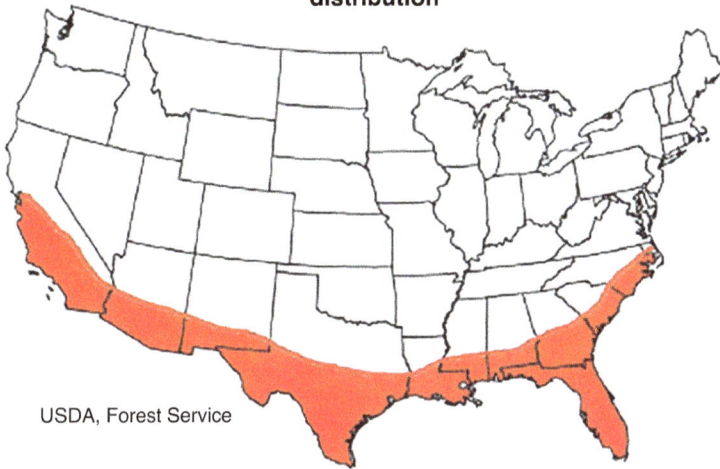

USDA, Forest Service

Fig. 6.11. Distribution of drywood termites in the USA. Credit: USDA

Fig. 6.12. Worker ant. Credit: Pixabay/CC0 Public Domain.

emerge. When a nest is disturbed, workers can be seen carrying away the pupal cocoons (which are often mistaken for eggs). Flying ants, which are the reproductives in the colony, emerge from the nest when the colony size enables a breeding cycle. At this event, winged female and male ants leave the colony to find a suitable nest location and mate; a new queen will establish the colony.

Trees and shrubs in green spaces or simply bare soil in the urban environment provide suitable nest sites. Ornamental plants and trees harbor the species of insects and other animals that serve as a food resource for ants. As aphids and other insects suck sap from plants, they excrete part of it as honeydew. This is a carbohydrate-rich food for ants. Species relying on honeydew as a food continue this feeding habit in natural and urban areas. Ants change their foraging to concentrate on food that is available or that fits the needs of the larvae in the colony. Worker ants do not eat solid food but crush it with their mandibles and take up any liquids

produced. Honeydew and other liquids are stored by the ant in a small stomach. This liquid can be regurgitated and fed to other workers, larvae, and the queen. Worker ants gather protein when larvae in the colony are developing. They forage for carbohydrates when larvae have completed development and use this as their own energy source.

Pavement ant

The common name fits the paved environment that this species occupies. The distinct mounds at the entrance to the nests is a common feature of sidewalks and green spaces in cities around the world (Fig. 6.14). The generalist feeding habits of these ants enable them to succeed in new environments. They are often the first animals to arrive at discarded food in cities. Urban populations have been found to show carbon-13 levels like those in human foods, suggesting that the pavement ant can utilize human food as a primary resource.

Pharaoh ant

The pharaoh ant is the most ubiquitous household ant in the world (Fig. 6.15). The workers are approximately 1.5–2 mm long. It is a major pest of residences, factories, office buildings, hospitals, and other areas where food is handled. Colonies have multiple queens and are divided

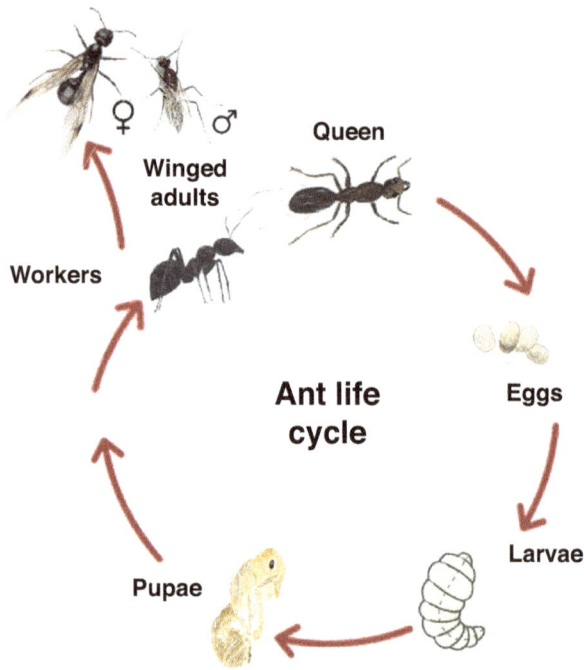

Fig. 6.13. Ant life cycle. Credit: W. Robinson.

Fig. 6.14. Pavement ant nest openings. Credit: Pixabay/CC0 Public Domain.

among many nest sites. They are constantly splitting into new colonies and nest sites within a structure. Part of the success and pest status of this ant relates to this fragmenting habit of the colonies. Mating takes place in the nest, and no swarms occur. Pharaoh ants use an attractant pheromone to establish a network of foraging trails, and another pheromone to mark food sources for nestmates to find. They also use a repellent pheromone to indicate a site or a food source that should be avoided by nestmates. Pharaoh ants are native to Africa. They are called pharaoh ants due to the mistaken belief that they were one of the plagues of ancient Egypt.

Fig. 6.15. Pharaoh ant worker. Credit: USDA.

Fig. 6.16. Argentine ant worker. Credit: USDA.

Argentine ant, or sugar ant

In the urban environment, nest sites of the Argentine ant include refuse piles, wall and masonry voids, masonry, and cracks in concrete around the perimeter of buildings (Fig. 6.16). Nests can be numerous in disturbed areas with permanent water supplies. Excessively dry or wet conditions often cause workers to invade houses. A network of interconnected nests usually exists, and workers from these nests regularly interchange among them. During the warm season, colonies may merge into supercolonies, in which the colonies in a large geographic region become one vast colony.

Their natural food is honeydew from aphids and mealybugs; indoors, they feed on sweets, meats, fruit, animal fats, and vegetable oils. Argentine ants produce a pheromone trail continuously when foraging and returning from a food source. Workers follow chemical rather than visual cues in the environment, which allows them to forage day and night. This species has an aggressive foraging behavior which often leads to the displacement of native ant species. The Argentine ant is successful in the urban biome because of its large colony size, foraging behavior, multiple queens, and because it is

Fig. 6.17. Spider webs on urban street lights. Credit: Pixabay/CC0 Public Domain.

attracted to disturbed habitats that are associated with human activity.

Spiders

Spiders have adapted to conditions around the outside and inside of structures in the human environment (Fig. 6.17). They occur in households around the world, whether urban or rural, regardless of the regional climate or cultural differences in the living space. The species that utilize household habitats may change, but they are almost always dependent on human activity for harborage and dispersal. The features of houses, such as the corners and overhangs adjacent to outdoor lights as well as undisturbed areas indoors, provide harborage and hunting grounds for these predators. Their food is the insects and other arthropods that live in household habitats or around the outside.

Pest status

Spiders may have been the first species to have pest status in household space. Spiders foraging on the ground for prey were an established part of the ecosystem when primitive house structures were built in open areas or on the forest edge. Once the opening to these structures was closed with a door, the importance of the space changed. Then it was separated from the outside, and entry was limited. Spiders would have easily entered the house at ground level while web-building species were also carried in on vegetation. The response to the first spider seen on a wall or on the floor would define their pest status. There would have been no apparent medical threat, as knowledge would have been limited; perhaps the response would have been only an aesthetic one regarding an unknown and unwanted intruder. Little has changed, as few spiders indoors are a medical threat to humans, but their presence is generally unwelcome in most cultures.

Foraging

Most spiders are general predators and prey on multiple species. Spiders have a low rate of metabolism, and they can store energy and go without feeding for long time periods. These biological traits enable spiders to live in the human environment. Some aspects of their biology provide an advantage in the long-term occupancy of the living space. Nocturnal foraging, small populations, and utilizing undisturbed areas allow limited interaction with people (Fig. 6.18).

There are relatively few spider species that have adapted to the urban environment to the extent that they are no longer represented by populations in the natural environment. There are few species that are dependent on humans for their habitats and dispersal. The spiders that seem adapted to the household environment are those in undisturbed areas such as cellars, basements, and storage areas. In these locations the temperature and relative humidity may be conducive for both the spider and the potential prey. The spider species that have shown some level of successful adaptation to indoor habitats include the common house spider, cellar spiders, and species of jumping spiders. They have a common occurrence indoors across rural and urban environments around the world (Fig. 6.19).

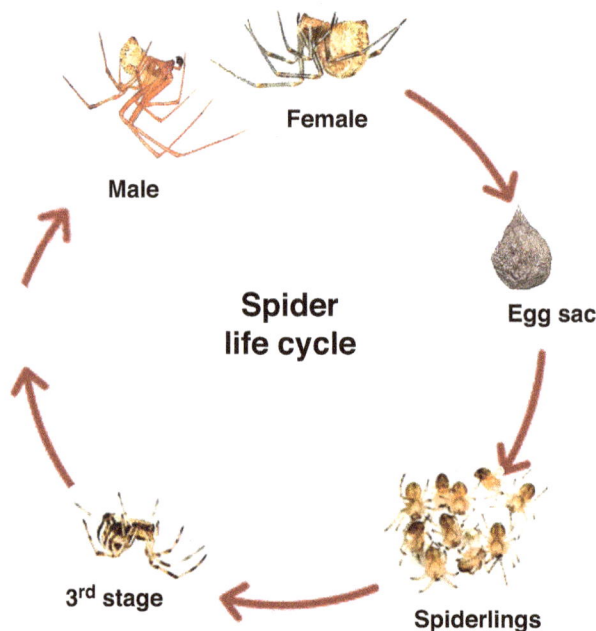

Fig. 6.18. House spider life cycle. Credit: W. Robinson.

Fig. 6.19. House spider. Credit: Pixabay/CC0 Public Domain.

Fig. 6.20. Wolf spider. Credit: Pixabay/CC0 Public Domain.

Wolf spiders

Wolf spiders have a wide global distribution. They are predators in outdoor habitats and do not establish populations indoors. These large spiders occur indoors in fall and spring (Fig. 6.20). When temperatures decrease in fall and increase in spring, the perimeter walls of buildings become "heat sinks" and are usually warmer than the air temperature. Wolf spiders detect the higher temperature and move toward it. They come inside along edges of windows and doors as they follow a temperature gradient. Wolf spiders do not build webs but instead hunt for their prey. Most species are active at night. The female carries the egg sac until it hatches, and the spiderlings cling to her back for a few days before dispersing.

Mediterranean brown recluse spider

The Mediterranean recluse spider has a global distribution due to human travel and the large amount of container shipping and regionally transported goods. This species has been introduced into parts of Europe, South-east Asia

from India to Japan, Australia, and Atlantic and Pacific islands. It occurs in basements and tunnels constructed in the urban biome. The webs function as a daytime shelter and seclusion for the egg sacs. Like other species in the genus, bites from this recluse spider can cause an area of permanently damaged cells that are slow to heal because of an enzyme (sphingomyelinase D) in the venom.

Brown recluse spider

The distribution of this species is primarily south-eastern USA, but it can be transported in household furniture.

Brown recluse spiders build irregular webs in woodpiles and sheds, closets, garages, and cellars. Human–recluse contact often occurs in isolated spaces when the spider is disturbed. They leave their webs at night to hunt. The body of both the male and the female is brown, except for a violin-shaped mark in the middle of the head region (cephalothorax) of the body (Fig. 6.21). The violin pattern is not a definitive identifier, as other spiders can have similar markings. Bites produce mild to severe pain. An open wound develops at the site of the bite, and

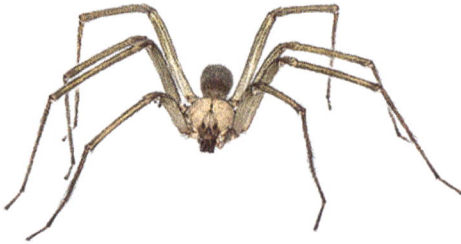

Fig. 6.21. Brown recluse spider. Credit: Pixabay/ CC0 Public Domain.

Fig. 6.22. Black widow (top) and redback spider (bottom) females. Credit: W. Robinson.

it may persist for several weeks. The bite may result in a lasting scar.

Black widow and redback spider

There are several species in the genus *Latrodectus* that occur as perimeter pests and indoors around the world. Species in Europe and South America are not known to occur indoors. The redback spider is distributed in South-east Asia, Australia, and New Zealand. Females are shiny black or brown, and the abdomen is rounded. There is a red mark, sometimes in the shape of an hourglass, on the underside of the black widow abdomen and a red stripe or two spots on the top of the redback spider (Fig. 6.22). The effects of a bite include increased heart rate and blood pressure, and often paralysis of the diaphragm muscles. Females do not move far from their nest web. Males are usually at the edge of the nest.

Green Bottle Fly

The green bottle fly occurs in urban and rural habitats in nearly all temperate and tropical regions. It is recognized by its metallic, blue-green coloration (Fig. 6.23). This fly species, *Lucilia sericata*, is important in forensic science. The presence of larvae is used to estimate the shortest time of the total post-mortem period. It is one of the first insects to arrive at a corpse, as it is attracted by chemicals produced at early stages of decomposition. Larval development time is linked to ambient temperatures.

Fig. 6.23. Green bottle fly male (top) and female (bottom). Credit: Pixabay/CC0 Public Domain.

Life cycle

Gravid females select the most suitable substrate for larval development by using semiochemicals to distinguish between fresh and aging carrion

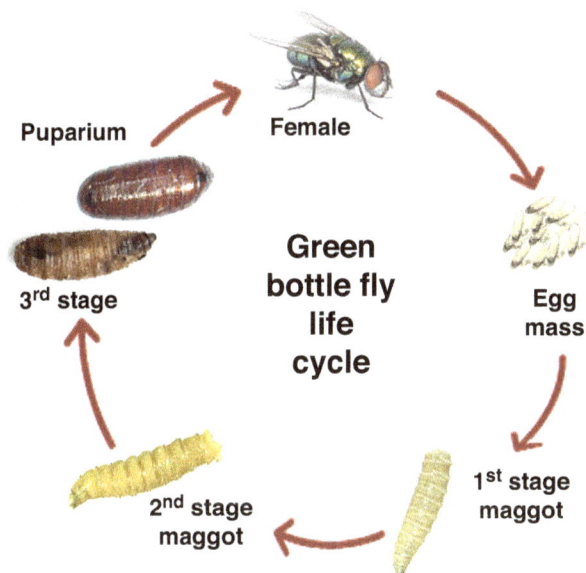

Fig. 6.24. Green bottle fly life cycle. Credit: W. Robinson.

before laying eggs (Fig. 6.24). This is done by orienting to dimethyl trisulfide emanating from fresh carrion. Large numbers of females may gather at a substrate that is optimal for egg laying (Fig. 6.25). Larval development is best in warm substrates, and females prefer to oviposit more eggs on warm substrates. Females lay a large mass of eggs at one time. There are about 200 eggs per mass, and females may produce a lifetime total of about 3000 eggs.

Larvae survival depends on the unique trait of feeding together in groups. The large egg mass leads to feeding of aggregates of same-age larvae. They benefit from the digestive byproducts of other larvae. Each larva secretes digestive enzymes and consumes the dissolved meat around it. Large numbers of larvae result in more digestive enzymes, which makes food more accessible for the whole group. At 27°C, the first larval stage lasts about 31 hours, then second about 12 hours, and the third about 40 hours. Third-stage larvae crawl from the substrate to a dry location to form the puparium.

Males can recognize flying females by the frequency at which the light from their iridescent body glints through their wings. They use the flashes to assess the age and sex of a potential mate.

Under direct sunlight there is a reflected flash at each wingbeat. Males recognize fertile females by this flashing light. They prefer a flashing on and off at 178 Hz. Sexually active males respond to a flash frequency from fertile females rather than being attracted by appearance or smell. Green bottle flies mate less frequently on cloudy days, since they rely on direct sunlight flashing through the wings of females to signal potential mates.

House Fly

The house fly has a cosmopolitan distribution and wide-ranging feeding habits (Fig. 6.26). It is attracted to animal and human wastes in the agricultural and urban environment. It can successfully adjust to available conditions by changing larval food preferences and adult dispersal to fit the environment. The house fly can complete multiple generations per year, and this enhances its genetic capability of developing resistance to insecticides.

Garbage that is scattered or contained in receptacles around houses and food service establishments can serve as a breeding site for house flies. These areas provide suitable egg-laying sites

Fig. 6.25. Green bottle fly females on substrate. Credit: Pixabay/CC0 Public Domain.

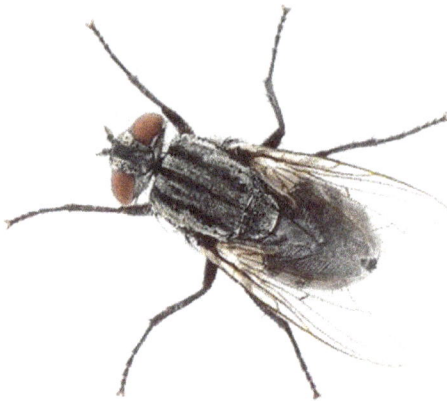

Fig. 6.26. House fly. Credit: Pixabay/CC0 Public Domain.

and food for the adults and maggots. The adult house fly is a carrier and vector of pathogenic organisms, such as the food-poisoning bacteria *Salmonella* and the bacteria *Shigella*, which causes diarrhea. The pest status of this insect is usually based on aesthetics but would be better placed on its potential effect on human health.

Life cycle

Adults disperse upwind and usually follow an odor stimulus. They search for food and water during the hottest and driest hours of the day; this is when adult activity peaks. They are inactive at night but move to artificial light during both day and night. Adult males and females feed several times per day on liquids or materials soluble in the salivary secretions from their mouthparts. Females generally mate only once and deposit their first egg batch within 5 days. Their potential egg production is about 900 in a lifetime. egg-laying females are attracted to warm and moist (40–70% water) organic substrates. A volatile semiochemical deposited by females on their eggs attracts other gravid females to the site, which results in large clusters of eggs. Development from egg to adult is about 7 days at 33°C. There can be 12 generations per summer.

Adult house flies are adapted to a liquid diet. Their mouthparts consist of a retractable, flexible proboscis with an enlarged and fleshy tip (Fig. 6.27). The tip is a sponge-like structure that

Fig. 6.27. House fly feeding on liquid. Credit: Pixabay/CC0 Public Domain.

Fig. 6.28. Red eye fruit fly. Credit: Pixabay/CC0 Public Domain.

has many grooves which enable the mouthparts to suck up fluids. House flies are often seen probing food surfaces with their mouthparts to obtain liquid food. Adults require several feedings of liquid food each day.

Fig. 6.29. Larva of red eye fruit fly showing extension of posterior spiracles. Credit: W. Robinson.

Synanthropy

The house fly is the only synanthropic, non-blood-feeding fly. The long synanthropy has allowed it to adapt to a variety of environmental conditions and food resources. A physiological connection to humans enables this species to taste and respond to lactose, which is the sugar found in milk. It helps to explain the association of house flies with lactating animals. This is a long-term relationship, as the house fly evolved as a species 86–48 MYA, while hominids emerged about 14 MYA. So, bottle flies, flesh flies, and the house fly group were present before the evolution of humans. All three are closely associated with the urban biome. The unique ability of the house fly to detect lactose indicates a link to humans and the domestication of lactating animals for milk.

Fruit Fly

These small flies are rarely recognized outside of buildings or laboratory colonies, but they are a part of the community of small flies that live in and around decomposing vegetation. They are feeding generalists that help to complete the decay process, but not to start it. Several species of fruit flies occur worldwide, and many have adapted to indoor habitats where fruit and vegetables are stored for long periods.

Red eye fruit fly

Adults of this common species have bright red eyes; the thorax and abdomen are yellowish brown (Fig. 6.28). Full-grown larvae are pale yellow, and the posterior spiracle (breathing) openings are brown and on a short extension of the larval body (Fig. 6.29).

Adults remain close to the breeding site or concentrations of fruits and vegetables, especially ripe fruit. eggs are laid in small groups directly on the substrate (Fig. 6.30). Hatching occurs in 24 hour and development time from egg to adult is 8–10 days at 25°C. Larvae often remain buried in the food substrate as they feed, with only their posterior spiracles exposed to the air. Third-stage larvae move from the substrate to a dry location to form the puparium. Adults emerge in about 4 days, and they mate and begin egg laying within 2 days. Adults live about 30 days in humid conditions. Outdoor populations are in garbage and decaying vegetable material. The red-eyed fruit fly colonizes rotting fruits at a particular time during the decay sequence. Other fruit fly species are attracted before this, followed by red-eyed fruit fly adults. This sequence is because the red-eyed fruit fly larvae have a higher tolerance to the ethanol in decaying fruit.

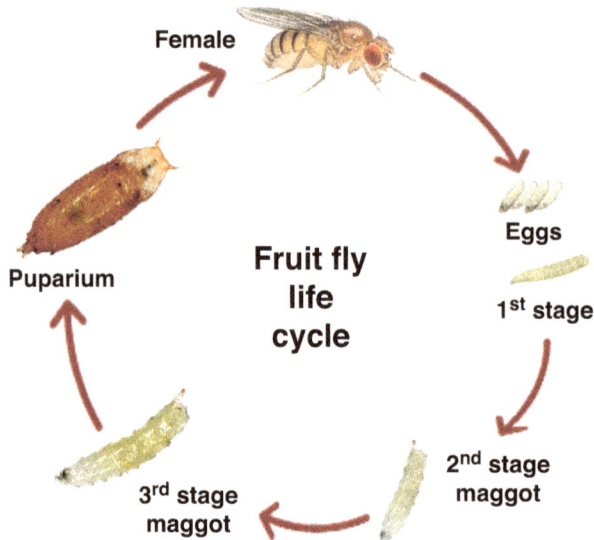

Fig. 6.30. Fruit fly life cycle. Credit: W. Robinson.

Dark eye fruit fly

Adults are brown to reddish brown; the abdomen is dark brown with yellow bands. Full-grown larvae are pale yellow, and the posterior spiracles are on an extension at the end of the larva body that is equal to the length of the last body segment. Adults are relatively inactive and rest away from the larval breeding site, which may be organic matter in floor drains or near garbage receptacles. eggs are laid on the substrate, and hatching occurs in about 24 hour. Larvae remain buried in the wet organic substrate, with only their posterior spiracles open at the surface. Development takes about 15 days at 25°C. This is a cosmopolitan species associated with decaying and fermenting organic material, including human feces, indoors and outdoors. It frequently occurs in food-processing and food-handling facilities and may infest these sites year-round.

Fig. 6.31. Phorid fly, *Megaselia scalaris*. Credit: W. Robinson.

the adults remain active nearly year-round. Adults are recognized by their zigzag running behavior on surfaces, and their appearance at windows. Adults are often active in locations that are a long distance from the site of the developing larvae. They will come to some traps that contain liquid for attracting fruit flies.

Larvae can develop on slightly decaying to putrefying organic material, which includes corpses of humans and other animals. The ability of the adult females to identify the early decay process allows larvae to be primary colonizers of decomposing material. This species often occurs in laboratory colonies of insects, such as crickets and cockroaches. Although the larvae have sclerotized (hardened) mouthparts

Phorid Fly

The phorid *Megaselia scalaris* probably has the highest level of fitness of any fly species in the urban biome (Fig. 6.31). Larvae of this species feed on a wide range of organic material, and

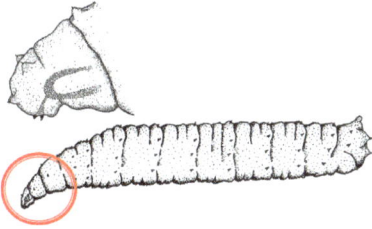

Fig. 6.32. Phorid fly larva (*Megaselia* sp.). Credit: W. Robinson.

Fig. 6.33. Phorid fly pupal cases before (left) and after (right) adult emergence. Credit: W. Robinson.

and can lacerate tissue, they are feeders on dead tissue (Fig. 6.32).

Females deposit eggs directly on substrates or on nearby surfaces. The features on the egg surface uniquely prepare the egg for a range of substrates. The dorsal surface is covered with sharp spines (spicula). The ventral surface appears smooth but is covered with small papillae. The dorsal spicula help the egg to mechanically attach to rough surfaces. The papillae on the ventral surface may help attach the e.g.g to hard, smooth surfaces.

Life cycle

The life cycle and duration of the immature stages is linked to an environmental temperature of about 25°C. The first and second larvae stages last 2 days, and the third for 4 days before pupation. The formation of pupal respiratory horns on the dorsal surface of the puparium signals the transition from the fourth-stage larva to the pupa. The pupal period lasts about 5 days. The life cycle can be completed in about 10 days. Adults emerge from the puparium by first dislodging the section with the respiratory horns (Fig. 6.33). Males develop and emerge several days before females.

The existence of reservoir populations of *M. scalaris* in the urban biome provides this scavenger with the opportunity to quickly find and utilize suitable substrates. The small (2.5 mm) body size and extendable ovipositor allows the female to deposit eggs in exposed and enclosed substrates. The larvae develop quickly and often give the appearance that the material was previously contaminated. During larval feeding and development, the substrate may become liquid or be exposed to water, and larvae are often in

Fig. 6.34. Brown rat (*Rattus norvegicus*). Credit: Pixabay/CC0 Public Domain.

semi-solid organic matter in drains. Phorid larvae do not have hydrophobic setae surrounding the posterior breathing spiracles as is the case in other flies (moth fly larvae) that feed in a liquid medium. The ability of larvae to take in air and create small bubbles to enable buoyancy may have larval survival value.

Brown Rat

The brown rat or Norway rat is successful in any climate zone and has infested urban biomes around the world. It adapts to conditions by relying on the physical strength and movement abilities of the adults, an omnivorous diet, and flexible nesting behavior. It is successful in urban habitats by utilizing the abundant harborage, kitchen waste, and discarded food. Urban garbage provides a more balanced diet for rats than discarded food in rural or farm areas. As a result, urban rats are usually larger than rats in other areas (Fig. 6.34). Urban green spaces provide a natural environment for digging burrows, and availability of human food waste. Urban sewer systems provide a consistent environment.

Life cycle

Female rats living in groups may have breeding synchrony. This can result in coordinated infestation peaks as large numbers of young enter the population at the same time. Among females that reproduce synchronously, approximately 80% of pups survive to weaning because there are more lactating females that can nurse young and there is less competition among pups. This phenomenon is typical in large and established populations (Fig. 6.35).

Rats reach sexual maturity at about 11 weeks, remain pregnant for about 24 days, and give birth to litters of 7 or 8ht pups. The young leave the nest and ingest solid foods at about 14 days after birth. There are up to 5 litters per year. Life expectancy is slightly more than 1 year. The ability to eat a range of food is a survival benefit for the brown rat. Adults have powerful jaws and hardened teeth. The large incisors have a Moh hardness scale number of 5.5, which is greater than concrete (5.0), plastic (3.5), and aluminum (2.0). The incisors are separated from the chewing

teeth, which permits the rat to gnaw on material without getting fragments into the mouth.

Harborage or burrow

Rats select habitats based on the availability of harborage, food, and water. Food availability determines the maximum number of animals that the habitat can sustain. Food and organic waste are an important food resource for rats. While food may determine the size of a rat population, the availability of harborage will determine whether a population is established. Brown rats tend to enter structures through sewage systems and cracks or holes in building foundations. Bown rats create their own harborage by burrowing into soil along a structure and establishing the space. The burrows will have more than one escape or exit, and the nest site will be well below ground (Fig. 6.36). Rats are less common in heavily built-up areas where there are mostly hard surfaces and little natural soil available for burrowing.

Home range and foraging

The size and shape of the home range is determined by access to feeding and harborage sites (Fig. 6.37). These associations lead to

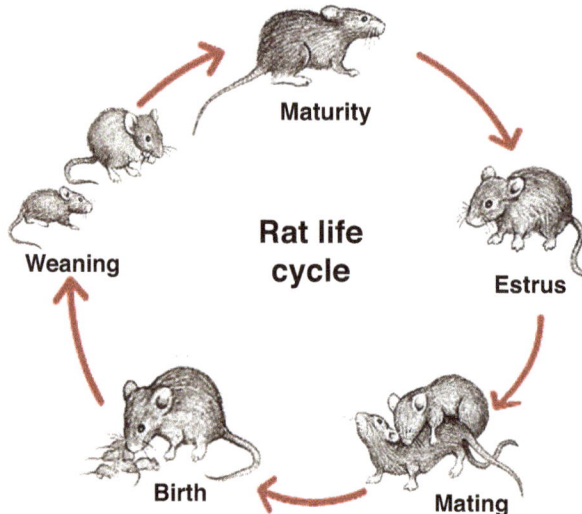

Fig. 6.35. Brown rat life cycle. Credit: W. Robinson.

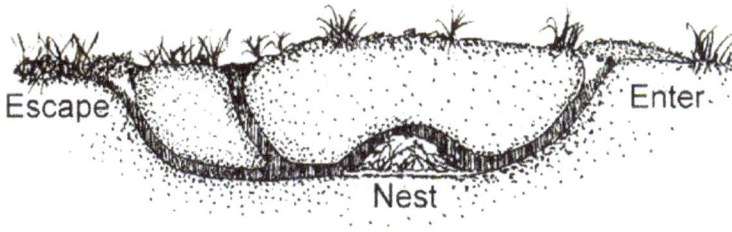

Fig. 6.36. Burrow of brown rat. Credit: W. Robinson.

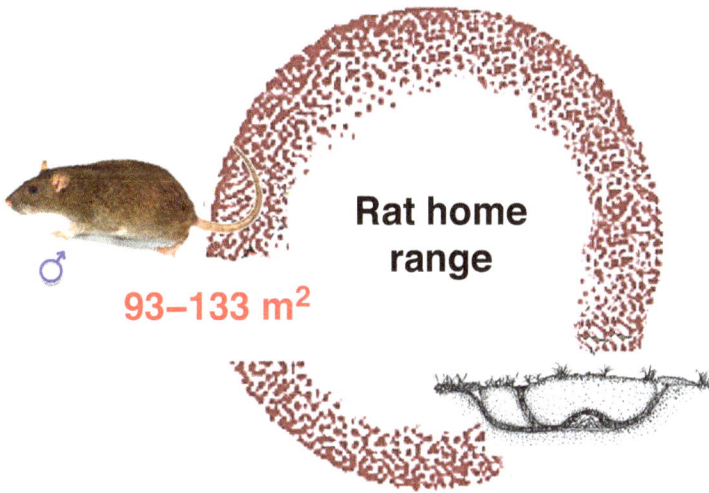

Fig. 6.37. Brown rat dispersal range. Credit: W. Robinson.

Rat home range

93–133 m^2

irregularly shaped home ranges and individuals moving along narrow pathways connecting harborage and food sources. Familiarity with the features of the home range may serve as a protective measure when individuals must quickly escape danger. Rats are familiar with their entire home range but spend about 50% of their activity in the core region. This familiar space includes access to food sources and is often in areas of dense vegetation. The survival value to individuals and an extended population is based on their use of the burrow for sleeping, eating and storing food, and raising young.

The brown rat can utilize a wide variety of plant and animal material as food, which means it can find suitable food in almost any urban habitat. Most of the food consumed is human food waste that is available within the home range of the reproductive female. Eating food in urban habitats has the potential for harm and a threat to survival if the food is toxic. The range of kitchen waste from modern households includes processed food and chemical additives that may affect young rats. Screening for harmful material while feeding young provides immediate and long-term survival. Fetal rats detect odor-bearing particles that come from their mother's diet and cross the placental barrier, and newborn rats respond positively to these foods. Nursing rats receive information about their mother's diet through her milk; they prefer the foods she ate during lactation.

Dispersal

Rat dispersal behavior is primarily linked to resource availability, competition, and mating behavior. When feeding and harborage sites are scarce, rats may travel significant distances in search of resources (Fig. 6.38). Mate-searching is an important driver of dispersal, particularly for males. Most migrants are reproductively mature males. Females move shorter distances than males. Rats are generally nocturnal, with

Fig. 6.38. Brown rats at storm sewer grate. Credit: Pixabay/CC0 Public Domain.

Fig. 6.39. House mouse. Credit: Pixabay/CC0 Public Domain.

heightened activity several hours before sunrise and after sunset. Males are generally more active than females. Brown rats move along the ground through narrow runways, near to fences and other cover. They travel between surface and sewer locations and between adjacent buildings. They do not travel between adjacent but separate sewer systems. The ability of rats to cross roadways depends on their width: larger roadways deter movement more than smaller roadways.

Control

Cats are commonly considered as a control tool for urban rats. This assumption is based on observations of domestic cats preying on mice. But there is limited evidence that feral cats can suppress established brown rat populations in cities. Feral cats prefer small (<250 g) prey such as mice, but brown rats in urban populations can be ten times heavier than mice. Research has shown that, instead of decreasing the number of rats in cities, cats primarily cause temporary space-use changes in rat behavior. Pressure from the foraging behavior of feral cats may be linked to cat populations and the availability of other food resources, including human food waste.

House Mouse

The house mouse is a species with great adaptability and behavioral flexibility that can be found in almost all urban environments (Fig. 6.39). The success of this small rodent is attributed to the ability to live in close association with humans and by using food and harborage in the living space. In urban housing, residents have ongoing infestations of house mice. Infestations also occur in food stores, warehouses, vacant buildings, empty lots, and garbage dumps.

Life cycle

Breeding occurs throughout the year in the urban environment, but wild populations have a reproductive season that extends only from April to September. Litters of five or six young are born about 21 days after mating. Young mice grow rapidly, and after 3 weeks they make short trips from the nest and eat solid food. They are weaned soon after and are sexually mature within 10 weeks of birth. A female may have ten litters per year, and this can result in indoor populations increasing rapidly when there is adequate food indoors.

Harborage and foraging

When living with humans, house mice nest behind rafters and in woodpiles, storage areas, or any secluded site near food. They construct nests from rags, paper, or other soft substances. They are generally nocturnal, although some are active during the day indoors. Adults are quick runners (up to 13 kph), good climbers and jumpers, and swim well. Despite this set of skills, they rarely travel more than 15 m from their foraging area.

House mice prefer to eat seeds and grain, but they will sample many kinds of items in their environment. Foods high in fat, protein, or sugar may be preferred even when grain and seed are present. A mouse consumes about one tenth of its body weight each day. Mice only need about 120 ml of water per day and they often lap drips from faucets in kitchens. They survive with little or no free water, as they obtain water from most of the food they eat. Absence of liquid water or food with low moisture content can reduce their breeding potential.

Dispersal

The forced dispersal of young mice in the colony contributes to the spread of infestations to adjacent areas. The spread in large apartment buildings is linked to movement of individual mice into neighboring units. Adjacent units with shared walls are likely to experience house mouse activity when one of the units has the initial infestation. Young male and female mice in the initial infestation follow walls, inter-apartment conduits, and odors to expand to new habitats. Pheromone-based scents and urine trails can guide mice to new areas in a habitat.

Communication and perception

House mice have excellent vision and hearing, a highly developed sense of smell, and their whisk-ers detect air movements and surfaces. They use pheromones and other odors to communicate with each other about social dominance, family composition, and reproductive readiness. Male mice produce ultrasonic sounds in response to female sex pheromones (Fig. 6.40).

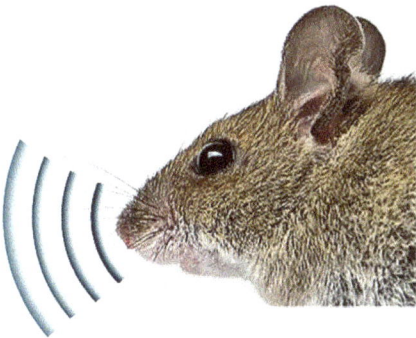

Fig. 6.40. Head of a house mouse. Credit: Pixabay/CC0 Public Domain.

Dust Mites

Several species of mites occur in the household. Two species are dominant because they feed on organisms associated with the human skin scales present in the living space (Fig. 6.41). A common feature of human dwellings is house dust. This is a thin layer of organic and inorganic material covering floors and furniture surfaces. The amount and spatial distribution is linked to activities of the human and animal inhabitants. The distribution of house dust mites is tied to the human skin scales in the dust.

Skin scales are microscopic particles pro-duced by the friction of clothing against the skin. They are shed from the skin and transmitted by air into a room. A healthy person sheds about 10 million skin scales daily. There are no apparent natural populations of these mites. They seem to be restricted to indoor, domestic habitats. Reservoir habitats are indoor sites with suitable humidity (75%) and temperature (27°C).

Life cycle

Development from e.gg to adult requires about 30 days. It progresses from a six-legged larval stage to nymph stages to the adult (Fig. 6.42).

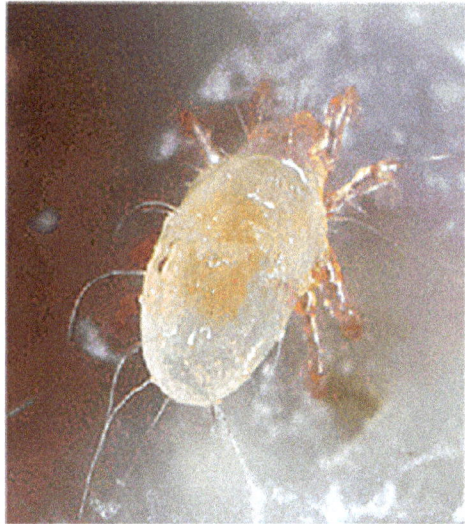

Fig. 6.41. Dust mite. Credit: Pixabay/CC0 Public Domain.

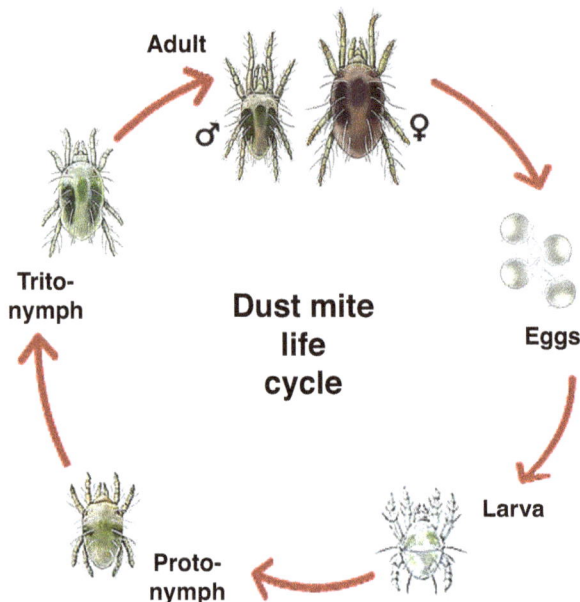

Fig. 6.42. Dust mite life cycle. Credit: W. Robinson.

Development time is influenced by the temperature and relative humidity in the immediate surroundings. Dust mites rely on atmospheric moisture as a source of body water, which makes the life cycle and population density determined by the household humidity. Development is accelerated by the warmth and humidity generated by individuals indoors. Body heat can produce temperatures as high as 35°C and up to 95% relative humidity.

resting behavior of the nymphs largely depends on aggregation pheromones contained in their feces. The pheromones seem to play a role in the induction of the quiescence. The prolonged quiescent protonymph is desiccation-resistant and this stage may extend several months before the next stage (tritonymph) develops. This later stage can survive dry periods, such as a long winter heating season when household relative humidity is low.

Development traits

In winter, when household temperatures are maintained at about 22°C, mite development is reduced. During summer, the warm and usually humid indoor conditions accelerate development. A significant dust mite adaptation to the household environment is a prolonged quiescent interval in the larval stage. Each life stage spends a percentage of time in a quiescent period, and it can be one-third to one-half of the total duration of the stage. This feature likely increases survival in an environment where there are seasonal and sometimes unpredictable fluctuations in conditions. The aggregation and

Habitat and substrate

Freshly shed skin scales are hard and dry, but after a few days on the carpet, furniture, or the bed pillow they absorb moisture and support the growth of mold. After the scales are attacked by mold, they are more acceptable as a food source for dust mites. Carpeting and cloth coverings on furniture provide a microclimate for fungi to attack the skin scales, and for mites to utilize both the fungi and the skin scales as food (Fig. 6.43). Fluctuating indoor temperature and relative humidity influence the conversion of dry skin scales into a suitable food and influence the water balance in mites. There is a limited number of

Fig. 6.43. Cross section graphic of pile carpeting. Credit: W. Robinson.

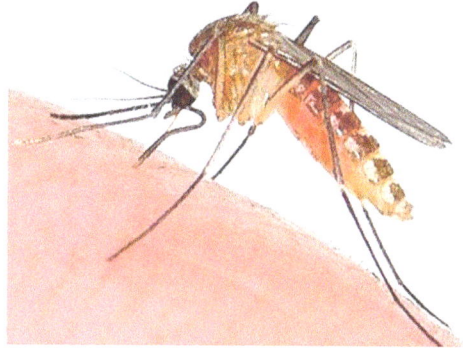

Fig. 6.44. *Culex* sp. mosquito. Credit: CDC US.

microhabitats in the household environment in which temperature and humidity are optimum for mite development, and these locations may be suitable on only a seasonal basis.

Distribution

In human dwellings, house dust mites are most prevalent in areas where people spend a large amount of time: beds, furniture, and tables. Carpeted floors support greater mite populations than do wood or tile floors. Carpeted floors around beds usually have a higher mite density than in other rooms. However, high mite densities also occur in cloth-covered furniture, and in the carpeting surrounding them. Cloth-covered furniture in a household is an important factor influencing mite density and distribution. Close contact between people and the fabric cover provides food and moisture required for mites present in the location. The greater the use of a piece of furniture, the greater the potential for high levels of dust mites on or around it. Frequently used chairs and couches in carpeted living rooms often support a high density of dust mites. Locations such as corners, the middle of rooms, and areas in front of doors where people walk and in front of windows usually have low densities of mites.

Dust mites can occur on the cloth seats of vehicles, and this may be a means of distribution of mites to locations outside the house, such as offices or other workspaces. Survival at these locations would depend on environmental conditions, but once established, mite populations would have the potential of an ongoing food resource.

Ecological trap

Household temperature and humidity and substrates such as floor and furniture coverings have continued to change with fashion trends and as construction materials and methods have improved. Many of these improvements, including central heating, wall and window coverings, and roofing materials, have provided for better control of indoor temperature and humidity. In temperate regions of the world, the indoor environment has become warmer during the cold months and cooler during the warm months. In household environments, there has been a reduction in the fluctuation of both temperature and humidity throughout the year. The humidity is the most important condition that determines the suitability of the indoor environment for insects such as dust mites and fleas. However, attaining a humidity that is suitable for humans and not for arthropod pests is difficult. In some regions, the use of carpeting as floor covering is decreasing, as it requires more care than hard surface material. The absence of the microclimate provided by carpeting may limit populations of dust mites.

Mosquitoes

Mosquitoes are important blood-feeding insects around the world (Fig. 6.44). They are vectors of several human diseases, including malaria, yellow fever, encephalitis, and dengue. Females of most species take a blood meal, while males feed on plant fluids. Larval stages are aquatic and occur in water in natural sites and in arti-ficial containers in urban habitats. The urban biome provides abundant sites for mosquitoes, and many species have adapted to the conditions and the available hosts.

Life cycle

Mosquito development is divided between larval stages that live in water and feed on microorganisms, and a flying adult female that feeds on blood, or a male that feeds on nectar (Fig. 6.45). Females are attracted to their host by exhaled carbon dioxide, body heat, and volatile chemicals on the skin of human hosts. Gravid females use mammal blood to form the eggs they lay. They can deposit up to 300 eggs at one time, but not all at one location. eggs hatch in about 7 days. Larvae complete development in about 14 days.

Breeding sites and dispersal

Egg-laying females prefer shaded sites with high water temperature and food for larval development. The type and distribution of landscaping in urban and suburban areas can influence the abundance of these sites and the mosquitoes

they host. Standing water is required for larval breeding (Fig. 6.46). Adult mosquitoes prefer to rest in vegetation where there is low light and limited air movement, both of which helps to prevent water loss. The more trees and vegetation that provide shade and wind-protected habitats, the greater the abundance of mosquitoes resting there. The same applies to larval breeding sites: shaded and shallow water are the best locations.

Female mosquitoes fly from their breeding site in search of a blood meal, but males usually remain close to the breeding site. Female flight range varies with species, time of year, and wind direction. Most adults disperse only 100–200 m from their emergence site.

Disease vectors

Mosquito-borne encephalitis viruses have become a major health concern in many regions of the world. Encephalitis is caused by bacterial and viral agents, and the resulting inflammation of the brain can result in high fever, seizures, prolonged illness, and sometimes death. These viruses are transmitted by mosquitoes feeding in bird and small animal populations, and include West Nile virus, which occurs throughout Europe, parts of Africa and Asia, and the

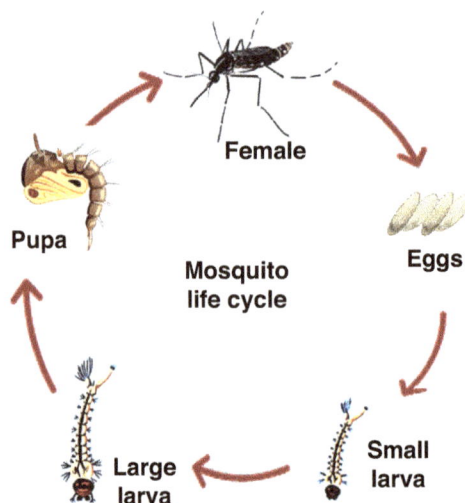

Fig. 6.45. Mosquito life cycle. Credit: CDC US.

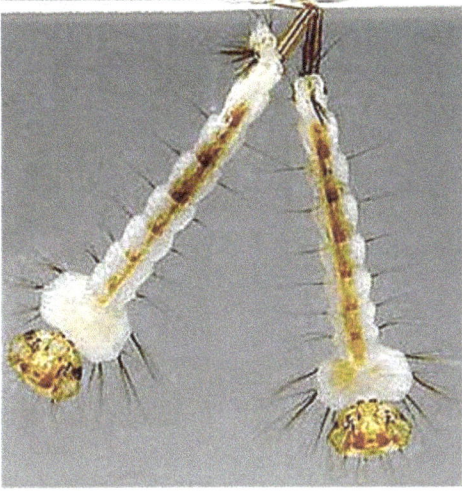

Fig. 6.46. Mosquito larvae (*Culex* sp.). Credit: CDC US.

Fig. 6.48. *Aedes aegypti* mosquito. Credit: CDC US.

Fig. 6.47. *Culex* sp. mosquito. Credit: CDC US.

USA, and Japanese encephalitis, which occurs throughout Asia to India.

Culex

These mosquitoes are distributed in urban environments around the world (Fig. 6.47). Foul water in street drains, gutters, and domestic containers and water with algae are preferred egg-laying sites. Larvae ingest diatoms, filamentous algae, bacteria, and other organic matter. Growth does not occur when the diet is dead organic matter.

Culex pipiens (complex) is the common house mosquito. It is in a large species-group consisting of several urban mosquito pests. Each species has adapted to a particular habitat or ecological zone. For example, *Cx. p. molestus* and *Cx. p. fatigans* only occur in urban environments, while *Cx. quinquefasciatus* (shown) is the southern house mosquito and occurs inside and breeds around the perimeter of houses.

Aedes

The biting of hosts occurs outdoors, and adults rest outdoors before and after feeding (Fig. 6.48). eggs are laid singly on a wet substrate, but they are resistant to drying for months. Winter is spent in the egg stage, and hatching occurs with spring rains. Ground pools are breeding sites for many species, but the common and important urban species use household containers holding water in urban habitats.

Aedes aegypti

This is the yellow fever mosquito (shown), and it is widely distributed throughout tropical and subtropical regions. This day-biting mosquito is adapted to living in the urban biome and is often found breeding and feeding around and in buildings.

Aedes albopictus

Adults fly close to the ground and do not fly in strong winds. Dispersal is 90–183 m from breeding sites. It is a day-biting mosquito, and there may be an early morning and late afternoon peak in biting. Females usually bite humans at ground level.

Anopheles

Anopheles mosquitoes are the primary vector of malaria around the world (Fig. 6.49). Larvae prefer unpolluted water. Adults are not strong flyers and are usually limited to short flights in low vegetation. Adults usually bite at night, inside or outside houses. Urban malaria is introduced from areas with high transmissions, as well as by the emergence of the urban malaria vector *Anopheles stephensi*.

Anopheles quadrimaculatus

This species is the most important vector of malaria in the eastern USA. Feeding occurs at

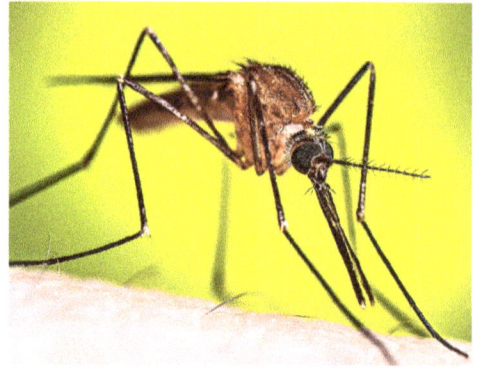

Fig. 6.49. *Anopheles quadrimaculatus* mosquito. Credit: CDC US.

night. During the day, the adults rest inside dark buildings and in dark corners. Flight activity peaks at dusk and a short period after dark, with limited flight for the remainder of the night, and at dusk they search for resting sites. Flight range is usually less than 2 km. Blood-feeding declines in fall and ends by November

Additional Reading

Dautel, H. and Kahl, O. (1999) Ticks (Acari: Ixodoidea) and their medical importance in the urban environment. In: Robinson, W.H., Rettich, F. and Rambo, G.W. (eds) *Proceedings of the Third International Conference on Urban Pests*, ICUP, Prague, Czech Republic, pp. 73–82. Available at: https://icup.org.uk/media/iimheito/icup481.pdf (accessed 29 September 2025).

Geier, M., Bosch, O., Steib, B., Rose, A. and Boeckh, J. (2002) Odour-guided host finding of mosquitoes: Identification of new attractants on human skin. In: *Proceedings of the Fourth International Conference on Urban Pests*, ICUP, Charleston, USA, pp. 37–46. Available at: https://icup.org.uk/media/3gnllmff/icup203.pdf (accessed 29 September 2025).

Krol, L., Langezaal, M., Budidarma, L., Wassenaar, D., Didaskalou, E.A. *et al.* (2024) Distribution of *Culex pipiens* life stages across urban green and grey spaces in Leiden, The Netherlands. *Parasite Vectors* 17, 37. DOI: 10.1186/s13071-024-06120-z.

Munshi-South, J. (2022) Urban rats and the smart city: A new paradigm for landscape-level rat management. In: Bueno-Marí, R., Montalvo, T. and Robinson, W.H. (eds) *Proceedings of the Tenth International Conference on Urban Pests*, ICUP, Barcelona, Spain, p. 20. Available at: https://icup.org.uk/media/ywqdpgod/3-munshi-south-plenary-pg-20.pdf (accessed 29 September 2025).

Murphy, R.G., Williams, H. and Hide, G. (2005) Population biology of the urban mouse (*Mus domesticus*) in the UK. In: Lee, C.-Y. and Robinson, W.H. (eds) *Proceedings of the Fifth International Conference on Urban Pests*, ICUP, Singapore, pp. 351–355. Available at: https://icup.org.uk/media/ihklchz5/icup054.pdf (accessed 29 September 2025).

Robinson, W.H. (2002) Role of reservoir habitats and populations in the urban environment. In: Jones, S., Zhai, J. and Robinson, W. (eds) *Proceedings of the Fourth International Conference on Urban Pests*, ICUP, Charleston, USA, pp. 217–223. Available at: https://icup.org.uk/media/aswjw2cb/icup223.pdf (accessed 29 September 2025).

Silverman, J. (2005) Why do certain ants thrive in the urban environment. In: Lee, C.-Y. and Robinson, W.H. (eds) *Proceedings of the Fifth International Conference on Urban Pests*, ICUP, Singapore, pp. 29–31. Available at: https://icup.org.uk/media/glclx2xa/icup004.pdf (accessed 29 September 2025).

Van Nes, A.M.T., Kort, H.S.M., Koren, L.G.H., Pernot, C.E.E., Schellen, H.L. *et al*. (1993) The abundance of house dust mites (Pyroglyphidae) in different home textiles in Europe, in relation to outdoor climates, heating and ventilation. In: Wildey, K.B. and Robinson, W.H. (eds) *Proceedings of the First International Conference on Urban Pests*, ICUP, Cambridge, UK, pp. 229–239. Available at: https://icup.org.uk/media/ty5djuri/icup632.pdf (accessed 29 September 2025).

Vargo, E.L., Perdereau, E., Dedeine, F. and Bagnères, A.-G. (2017) Global invasion history of the termite *Reticulitermes flavipes* (Isoptera: Rhinotermitidae) as revealed by three classes of molecular markers. In: Davies, M.P., Pfeiffer, C. and Robinson, W.H. (eds) *Proceedings of the Ninth International Conference on Urban Pests*, ICUP, Birmingham, UK, pp. 129–131. Available at: https://icup.org.uk/media/uihbyalh/icup1194.pdf (accessed 29 September 2025).

7 House as Ecosystem

The structural house is a dominant feature of the urban biome. A house can be considered an ecosystem. It is a space where living organisms, such as people, pets, insects, and plants, interact with the non-living components, such as the structure, floor covering, fabric, and furnishings (Fig. 7.1). These elements interact with each other in a complex of microhabitats in a protected and environmentally controlled space. Some of the house substrates are a food resource or harborage for species that live in them or feed on them.

There is a level of uniformity in the "house" structural features and activities across cultures and continents. There are basic environmental components of human habitats that are required for a safe and healthy living space. Cultural habits and everyday life are expressed in the spatial layout, the furnishings, and the basic activities of cooking, eating, and sleeping. Modern societies have brought plants and animals to the living space. These additions influence the environmental conditions and interactions with humans in the space. The urban biome will develop with global urbanization and with each generation of people living and shaping it to their needs.

As urbanization will bring new housing units and efficient construction methods, existing units will be modernized as demand increases. The most likely change accompanying urbanization will be a reduction in fluctuations in temperature and humidity. Controlling these two conditions will be critical as climate change will continue to challenge urban inhabitants. An early house construction goal in temperate regions was to reduce heat loss, but climate change has created the issue of house overheating. This earlier goal of heat conservation

Fig. 7.1. House floor plan. Credit: W. Robinson.

DOI: 10.1079/9781800626416.0007

has become a structural trap that is difficult to redress. Studies of indoor activity patterns report that individuals in some European countries spend 66% of their time indoors, and people with compromised health may spend even more time indoors.

The influence of an overheated house ecosystem on the various species inside may be linked to the basis of their synanthropy. If the temperature or humidity exceeds their range of survival, some species may have limited options to adjust. Remaining indoors to maintain or establish reproductive populations would be a genetic and/or epigenic challenge. Increased information on the indoor biome species will change the already known interaction of the triangle of species–harborage–resource and uncover new opportunities to reduce pest populations and pest status.

Grain, Flour, and Carpet Beetles

There are several species of beetles associated with stored food in the household environment. The majority of these are general feeders. They may be scavengers on small amounts of food, but large populations develop when large quantities of cereals and breakfast foods are infested.

Granary weevil

There is historical evidence of the association of this species with early Neolithic agriculture. The granary weevil would feed on most of the primitive grains grown and stored in early agriculture, including wheat, barley, oats, corn, and rice. The adaptations of this synanthropic grain pest include larval development inside the grain kernel to limit exposure to low humidity, and wingless adults, helping to limit water loss.

Adults are about 5 mm long and reddish-brown; neither males nor females are capable of flight (Fig. 7.2). Females lay about 250 eggs; a single egg is deposited in a grain kernel. Females first chew a hole in the kernel, deposit an egg, and then seal the hole with a secretion. All larval stages and the pupal stage occur within the kernel; after development the adult chews a hole

Fig. 7.2. Granary weevil adult. Credit: USDA.

Fig. 7.3. Sawtoothed grain beetle. Credit: USDA.

in the kernel and emerges. The life cycle takes about 5 weeks in the summer but may take up to 20 weeks in cooler temperatures. Adults also feed on grain kernels, and they live up to 1 year.

Sawtoothed grain beetle

These beetles are small (2.5 mm long) and have a flattened body (Fig. 7.3). The adults are capable of penetrating food packages and living in narrow crevices in household cabinets. Infestations also occur in stored grain. Although they do not fly, the adults are transported with food packages. The name is based on the

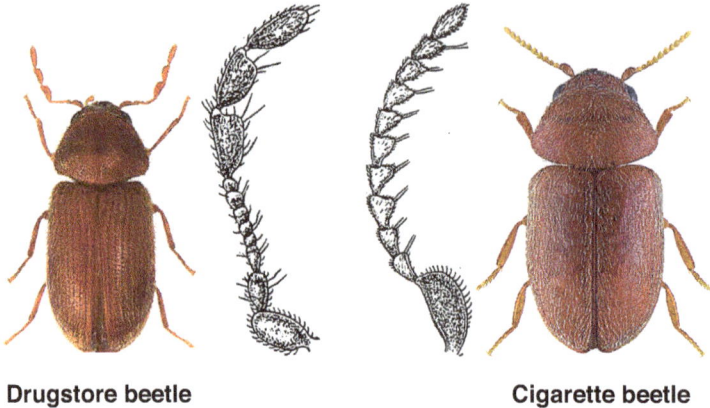

Fig. 7.4. Cigarette beetle and drugstore beetle. Credit: USDA.

Drugstore beetle   **Cigarette beetle**

sawtooth-like projections on each side of the thorax.

The average number of eggs per female is 280. eggs are laid on the surface or in crevices in bulk foods. eggs hatch in about 16 days; there are the larval stages. In the third stage, a pupal chamber is constructed by cementing together particles of food. Development from egg to adult takes about 80 days. The female lifespan is about 12 weeks.

Cigarette beetle and drugstore beetle

The cigarette beetle and the drugstore beetle (Fig. 7.4) feed as larvae on coffee beans, dried fruit, wheat flour, herbs, and leather. Adult cigarette beetles feed on the same food as the larvae. Adult drugstore beetles do not feed.

The egg-laying period is about 20 days and is influenced by temperature and availability of water. There are six larval stages. The lowest humidity for development is about 25% for the cigarette beetle and 35% for the drugstore beetle. The length of the development period varies with the food source. The last larval instar constructs a pupal chamber from the food substrate. The pupal period lasts about 12 days.

Varied carpet beetle and furniture carpet beetle

Larvae of carpet beetles feed on seeds, grains, cereals, wool rugs, blankets, clothing, silk, furs, feathers, and dead animals. During development

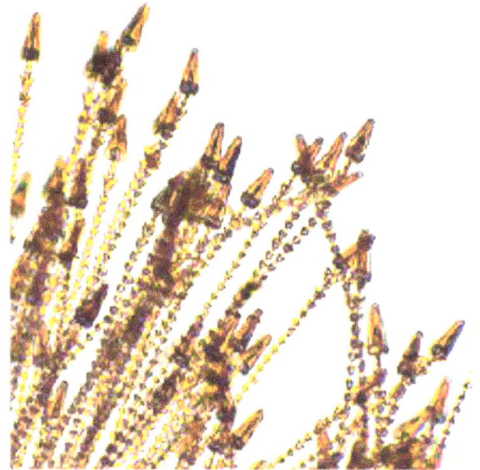

Fig. 7.5. Hastisetae setae. Credit: USDA.

from one larval stage to the next, larvae leave behind their previous skins on the infested material. They often feed in a small area, and their cast skins accumulate in that location. These are usually the only indication of an infestation, as the adults are rarely seen. They may collect at windows or move outdoors to feed on the pollen of flowers.

Hastisetae (*hasti*, Latin for spear-shaped) are detachable setae on the thorax and abdominal segments of larvae (Fig. 7.5). They are primarily used for defense against invertebrate predators. They entangle the long hairs, the antennae, and the legs of an attacking predator, which enables the larva to survive. However, hastisetae from infesting larvae can contaminate stored food

Fig. 7.6. Varied carpet beetle larva and adult (top), furniture carpet beetle larva and adult (bottom). Credit: USDA.

and fabrics, and they can cause an allergic reaction and threaten human health.

Varied carpet beetle adults and larvae (Fig. 7.6, top) are often found indoors on the remains of dead animals in wall voids. The larvae are elongated and densely covered in long setae. These setae are in alternating rows of light and dark brown patches: the larva appears covered in brown stripes.

Furniture carpet beetle larvae (Fig. 7.6, bottom) feed on animal tissues and products based on them, including hair, wool, skin, and bone. This beetle is also a pest in museums, where it attacks dead insects, and in libraries, where it attacks leather bindings.

Red flour beetle and confused flour beetle

The red flour beetle (Fig. 7.7) infests stored grain, flour, cereals, pasta, biscuits, beans, and nuts. This species and the closely related confused flour

Fig. 7.7. Red flour beetle. Credit: USDA.

beetle are the two most common pests of stored food commodities worldwide. They are successful in heated environments, and the adults some- times can live more than 3 years. The physical difference between the two species is the shape of their antennae. The confused flour beetle increases gradually in size; the antennae of the red flour beetle have three large segments at the end. Red flour beetles can fly short distances, but confused flour beetles do not fly.

These flour beetles are often found in agri- culturally stored grain and in household flour, dried fruit, and nuts. Stored food is damaged by beetle feeding and by the accumulation of their dead bodies, fecal pellets, and foul-smelling secretions. The red flour beetle has developed resistance to organophosphates, carbamates, cyclodienes, pyrethroids, and some fumigants, which makes control difficult. The two species have very different susceptibilities to insecti- cides: populations of confused flour beetles are generally less resistant than those of red flour beetles.

Flour Moths

Species that consume flour products and natural fabric material are feeding directly on and in their harborage. Their life cycle centers on the survival of the immature rather than the adult stage. Larvae of these species use silk to condition their harborage and to make a pupal cocoon. Clothes moth immatures show a similar behavior but are either enclosed in a silk case or remain in a silk retreat during development.

Indian meal moth

The Indian meal moth (Fig. 7.8) is an impor- tant stored product pest of the post-harvest stages of grains. It has a global distribution and is the most common stored product moth species. The larval stage feeds on and damages grain and grain products, dried fruits, and flour products. Adult moths do not feed. Populations of this species have developed physiological resistance to a range of chemical and biological insecticides.

Life cycle

Soon after mating, the female moth begins laying about 300 eggs, singly or in groups, directly on food material. Hatching occurs in 3 days, and development is completed in about 3 weeks. The fifth larval stage moves away from the larval food before pupating (Fig. 7.9). Diapause is a common feature in the life cycle of species that infest stored grains, including the Indian meal moth. It is a period of 2–9 months in which all feeding activity and devel- opment is suspended. It occurs during the last larval period.

Diapause

This condition is induced by photoperiod length, temperature conditions, or larval food quality. It is an important adaptive feature that is linked to the success of the Indian meal moth in stored food. Diapause may be triggered by decreas- ing day length. It has also been found that temperatures of about 25°C when the daylight period is ≤13 hour cause rapid diapause induc- tion. Photoperiodism plays a primary role in regulating the induction of diapause since it is used almost universally by insects in different geographical areas. The adaptive values of dia- pause are that it occurs only during the larval stage, and that it may be either facultative, as in response to environmental cues, or obligate, as occurring in each generation regardless of environmental cues.

Insect development indoors depends on envi- ronmental conditions and the food resource. In the case of the Indian meal moth, the larval food

Fig. 7.8. Indian meal moth adult. Credit: USDA.

Fig. 7.10. Casemaking clothes moth in case. Credit: USDA.

Fig. 7.11. Casemaking clothes moth larva and opened case. Credit: USDA.

Fig. 7.9. Indian meal moth larva. Credit: USDA.

resource is also the larval harborage. Facultative diapause is critical to the survival of the Indian meal moth because it can respond to unfavorable resources or environmental conditions. It can synchronize the life cycle with favorable conditions. The long-term success of this species is, in part, linked to its ability to respond to unfavorable conditions with a halt in development by a large percentage of the population.

Clothes Moths

The primary pests of natural fabrics in the house habitat are two species of moths: the webbing clothes moth, which makes a silk case that attaches the larva to the substrate, and the casemaking clothes moth, which is enclosed in the silk case it carries (Fig. 7.10). The webbing clothes moth is tolerant of dry indoor environments, and its economic importance increased with the use of central heating systems.

Life cycle

The small (8–12 mm long) adult moths are weak flyers and are not attracted to artificial lights. The female prefers to walk on surfaces rather than fly when laying eggs. The males actively search for females in the infested fabric. eggs are attached to fabric threads, and laying about 40 eggs can extend over 2 weeks. eggs hatch in about 10 days. The larva of the webbing moth constructs a tunnel of silk around itself and feeds on the substrate below it. The casemaking moth larva spins a silk case and carries the case with it as it feeds (Fig. 7.11). Casemaking moth larvae complete growth in 3 months, but it can be 36 months for the webbing moth. Development can extend to years under unfavorable conditions.

Webbing and case-making moths can coexist when household microhabitats have different humidity levels. Both species prefer to feed on portions of fabric soiled by human sweat, which satisfies their need for vitamin B. Larvae are attracted to common stains, such as human sweat, human urine, and fruit juices. However, clean woolens can be also attacked. The nutritional flexibility of the webbing moth is an adaptive advantage for modern household conditions. In addition to plant-based material, pure plant-based material such as cotton cannot be digested by either species. Larval stages can eat and digest keratin-based material, such as woolen textiles and carpets, blankets, and clothing. This level of fitness increases dispersal potential to microhabitats and persistence in the urban biome.

The silk case around the larval stages of both species may limit water loss and increase survival in low-humidity conditions. The case of the webbing moth is on the fabric surface, which may provide additional survival value for the habitat. The larva of the webbing moth has a sclerotized head capsule, but the dorsal surface of the thoracic segment behind the head is only slightly sclerotized (Fig. 7.12). Larval feeding and development primarily occurs inside the protective case, and the body segments have little exposure. The head and first segment of the thorax of the casemaking moth larva are well-sclerotized and dark colored. This feature may limit water loss as these segments remain extended from the feeding tube when feeding and moving.

Fig. 7.12. Webbing clothes moth larva head and thorax. Credit: USDA.

structural framing, and furniture. Both Lyctids and Ptinids exist in natural populations. Development indoors is linked to the moisture content and carbohydrate content of the wood. Larvae are adapted to initial low moisture content and fluctuations after entering and feeding.

Wood-infesting beetles are morphologically and physiologically adapted to using wood as a food source. Larval mouthparts can tear and remove pieces of wood for digestion. The *"Ptinid* strategy" of grinding food and utilizing cellulose as a source of carbohydrates provides these beetles access to a variety of woods for food. In the urban environment, Ptinids infest structural softwoods and hardwoods in attics, crawl spaces, and living spaces. The *"Lyctus* strategy" utilizes hardwood species with a high nutrient value. The female selects a larval substrate by first chewing a small amount to determine that the carbohydrate content is suitable for larval development.

Powderpost Beetles

Beetles in the families Ptinidae (previously Anobiidae) and Lyctidae are important pests of structural timber and are usually called powderpost beetles. Larvae of Lyctid beetles infest seasoned hardwoods. In the urban biome, they infest stored lumber, interior woodwork, and hardwood furniture. In the household environment, these beetles infest flooring,

Lyctid beetle life cycle

Females preparing to lay eggs chew a small amount of the wood to determine that the carbohydrate content is above 3% before depositing eggs. eggs are placed in pores of the wood. The larval stage lasts about 9 months, but development may be extended to 2 or more years as the wood ages. Full-grown larvae tunnel close to the wood surface and form a pupal chamber below; the adult cuts the exit hole to emerge. Lyctids are most active in wood with a moisture content of 10–20%. In air-dried lumber, much of the starch content is degraded, but after kiln drying, the starch content is fixed and the wood is susceptible to attack for many years.

Ptinid beetle life cycle

Females lay eggs on the wood surface; they prefer wood that is 2–5 years old and with a rough surface. The first-stage larvae bore into the wood soon after they hatch. Larvae stop or reduce feeding in response to low (winter) temperatures, and resume feeding in spring. Full-grown larvae tunnel close to the surface and prepare a pupal chamber. Adults emerge in spring and mid-summer (Fig. 7.13). Emergence holes of the adults of wasp parasites of the larvae may be present in the wood surface (Fig. 7.14).

Ptinids populations grow slowly, and many do not produce structural damage. A large infestation is needed to cause a critical 24% loss in the elasticity of structural pine wood. This level of infestation may require 10 or more years and very favorable environmental conditions.

Carpenter Ants

Camponotus is the largest ant genus with about 1000 species, and they are distributed in tropical and temperate regions. The common name for these large ants comes from the habit of many species for nesting in wood, and for the nest galleries being sandpaper smooth. Although nearly all species nest in wood, they do not feed on wood. Workers are scavengers on animal materials in natural and domestic habitats. The body color of adults and workers ranges from black to reddish black (Fig. 7.15).

Wood-infesting ants in the genus *Camponotus* are morphologically adapted to using structural wood as a nest site. The large and small workers in the colony have a broad head that accommodates musculature to operate the large, toothed mandibles (Fig. 7.16). These mouthparts can tear and remove pieces of moisture-damaged and sound wood for excavating nest galleries. This process results in fibrous frass and smooth surface channels. The sharp teeth on the mandibles of the young workers are fit for nest creation and expansion, but as the mandible teeth wear, the task of the workers changes to fit the condition. The *"Camponotus* strategy"

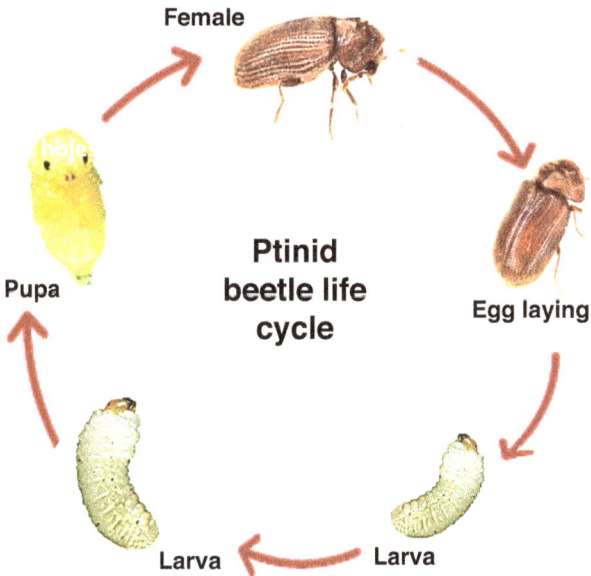

Fig. 7.13. Ptinid (Ptinidae) beetle life cycle. Credit: W. Robinson.

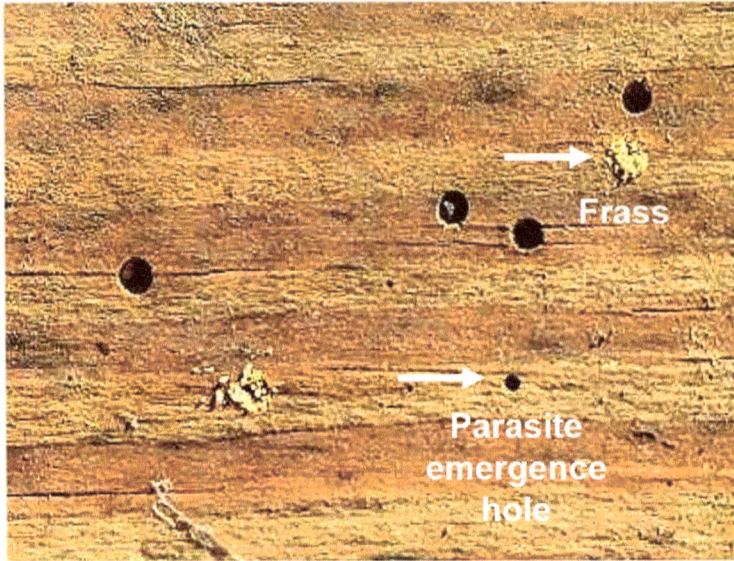

Fig. 7.14. Ptinid beetle and larval parasite emergence holes in wood. Credit: W. Robinson.

uses the fitness of the mandibles throughout their gradual morphological transition. This strategy determines the task and the tool (mandibles) of the workers to benefit the overall colony.

Mandibular teeth wear

The gradual wear of the mandibles of the worker ants begins with only slight wear of the incisor teeth (Fig. 7.17). The next stage shows wearing of all teeth, and many have rounded or broken tips. Finally, the mandible shows an absence of teeth, and the apical incisor is absent or small. The large incisor is often the last tooth to show significant wear. It may have nearly twice the hardness of the other teeth and be strengthened with zinc (Zn). Instead of wear resulting in reduced utility of workers, it may increase foraging efficacy when they transition from being inside the nest to outside. When wear reaches the point that the teeth are nearly gone, the mandibles may be better suited for dissecting insect prey than gouging wood in galleries. While foraging on the ground or in trees, they may encounter disabled insects and can carve out pieces of flesh to be returned to the nest.

Pest status is based on the structural damage and household nuisance caused by several species in the urban environment. In northern USA and southern Canada, carpenter ants replace subterranean termites as the most important wood-infesting insect in the urban environment. Nests are in live or dead trees, in rotting logs or stumps, or in sound or moisture-damaged structural wood. Wood is excavated by the workers with their large mandibles, and the fibrous frass is removed and pushed out of the galleries or packed into unused tunnels.

Carpenter ant galleries are distinguished by their smoothness and irregular shape (Fig. 7.18). They usually do not follow the natural grain of the infested wood. The frass excavated from the galleries is packed into unused galleries or discarded from the galleries. It collects in small piles, often of long fibrous strips. Galleries usually start in moisture-damaged wood but can be extended in adjacent sound or undamaged wood. The damaged wood can become structurally weakened and not able to support weight.

Colonies are naturally large and usually divided into small satellites and a parent colony in the main nest site. The parent colony is usually in a humid environment and contains the functional queens and developing brood, including the eggs and small larvae. At the satellites there are only

Fig. 7.15. Carpenter ant queen without wings. Credit: Pixabay/CC0 Public Domain.

Fig. 7.16. Head of carpenter ant showing mandibles. Credit: W. Robinson.

Fig. 7.17. Carpenter ant mandibles showing wear of the teeth. Credit: W. Robinson.

workers, full-grown larvae, pupae, and sometimes winged reproductives.

Large workers have heads disproportionately larger than those of small workers. The age of a colony is indicated by the proportion of large workers in the worker population. The percentage of large workers is high in old colonies and low in young colonies. The total number of workers in a colony when reproductives are produced may be up to 12,000.

Natural food for carpenter ants and most other ant species includes honeydew, plant exudates, and live and dead insects and other arthropods. They also scavenge on the carcasses of dead vertebrates. Indoors they forage for sweets and high-protein foods. Most species of carpenter ant forage at night and rely on a trail pheromone for orientation between the nest and

Fig. 7.18. Carpenter ant-damaged wood. Credit: W. Robinson.

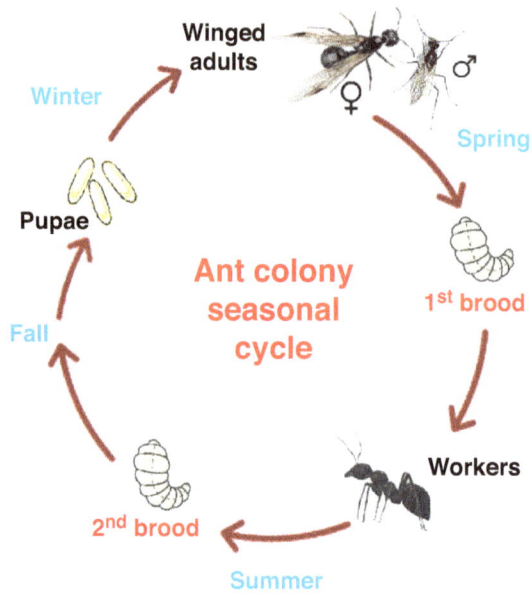

Fig. 7.19. Ant colony seasonal cycle. Credit: W. Robinson.

a food source. They also orient to the sun, the moon, or street lights at night. During spring and early summer, when brood production is at a peak, the workers forage for proteins, which are fed to developing larvae (Fig. 7.19). In late summer and fall, workers change their foraging habit to carbohydrates, which are used as an energy source by adults.

Additional Reading

Bertone, M.A., Leong, M., Bayless, K.M., Dunn, R.R. and Trautwein, M.D. (2017) Indoor arthropod communities and distributions in U.S. homes. In: Davies, M.P., Pfeiffer, C. and Robinson, W.H. (eds) *Proceedings of the Ninth International Conference on Urban Pests*, ICUP, Birmingham, UK, pp. 17–23. Available at: https://icup.org.uk/media/r31bgbrw/icup1178.pdf (accessed 29 September 2025).

Hansen, L.D. and Akre, R.D. (1993) Urban pest management of carpenter ants. In: Wildey, K.B. and Robinson, W.H. (eds) *Proceedings of the First International Conference on Urban Pests*, ICUP, Cambridge, UK, pp. 271–279. Available at: https://icup.org.uk/media/q1mntwta/icup638.pdf (accessed 29 September 2025).

Juson, A. and Juson, C. (2017) Ability of trained scent detection dogs to detect grain weevil in wheat samples. In: Davies, M.P., Pfeiffer, C. and Robinson, W.H. (eds) *Proceedings of the Ninth International Conference on Urban Pests*, ICUP, Birmingham, UK, p. 471. Available at: https://icup.org.uk/media/gm4jr1md/icup1273.pdf (accessed 29 September 2025).

Mináø, J. (1999) Synanthropisation and spreading of Dermestidae (Insecta: Coleoptera). In: Robinson, W.H., Rettich, F. and Rambo, G.W. (eds) *Proceedings of the Third International Conference on Urban Pests*, ICUP, Prague, Czech Republic, p. 675. Available at: https://icup.org.uk/media/4mud2plg/icup547.pdf (accessed 29 September 2025).

Plarre, R. (2014) Likelihood of infestations by *Tineola bisselliella* (Lepidoptera: Tineidae) from natural reservoirs. In: Müller, G., Pospischil, R. and Robinson, W.H. (eds) *Proceedings of the Eighth International Conference on Urban Pests*, ICUP, Zurich, Switzerland, pp. 345–352. Available at: https://icup.org.uk/media/sssikro1/icup1135.pdf (accessed 29 September 2025).

Plarre, R., Hering, H. and Matzke, M. (2017) Parasitoids for classical biological control of *Tineola bisselliella* (Lepidoptera: Tineidae). In: Davies, M.P., Pfeiffer, C. and Robinson, W.H. (eds) *Proceedings of the Ninth International Conference on Urban Pests*, ICUP, Birmingham, UK, pp. 335–345. Available at: https://icup.org.uk/media/s1vljblc/icup1228.pdf (accessed 29 September 2025).

Plarre, R., Busweiler, S., Haustein, V., Von Laar, C. and Haustein, T. (2022) *Korynetes caeruleus* (Coleoptera: Cleridae) for biological control of *Anobium punctatum* (Coleoptera, Ptinidae). In: Bueno-Marí, R., Montalvo, T. and Robinson, W.H. (eds) *Proceedings of the Tenth International Conference on Urban Pests*, ICUP, Barcelona, Spain, pp. 34–44. Available at: https://icup.org.uk/media/lyipplv5/6-plare-1-f-pp-34-44.pdf (accessed 29 September 2025).

Zorzenon, F.J., de Carvalho Campos, A.E., Junior, J.J. and Potenza, M.R. (2011) Survey and management of carpenter ants on urban trees in the city of São Paulo, Brazil. In: Robinson, W.H. and de Carvalho Campos, A.E. (eds) *Proceedings of the Seventh International Conference on Urban Pests*, ICUP, Ouro Preto, Brazil, pp. 73–76. Available at: https://icup.org.uk/media/a44pp22l/icup1080.pdf (accessed 29 September 2025).

8 Synanthropic Traps

Synanthropy is a dynamic condition established between the organism (synanthrope) and the habitat. Synanthropes have developed morphological, physiological, and life cycle adaptions that enable populations to persist in the built environment (Fig. 8.1). The adaptions are linked to conditions in the habitat they occupy. Their long-term presence depends on their ability to adjust to changes, especially those created by human activity. However, for some species the change exceeds their ability to quickly respond.

The concept of a synanthrope is based on the principle of adaptions created over time in a natural habitat and employed in a new habitat in the urban biome. Synanthropes have a set of traits that fit a specific habitat or niche. Synanthropic species are closely integrated into the habitat and usually are not passive inhabitants. They shape the habitat and influence conditions, especially in harborages, for their benefit and survival. But conditions in the urban biome undergo steady change, and human decisions become drivers of change. Cultural influence may induce gradual changes in features, while climate and economics may induce abrupt and permanent changes.

The term "synanthropic trap" describes an outcome of environmental change, and the concept identifies an end condition for some species in the urban biome. Some environmental changes may position synanthropes at the limit of their temperature range, food resources, or means of dispersal. If species populations are unable to adjust, a synanthropic trap may be engaged. In this condition, the original environmental conditions have changed beyond the species survival range.

Trap Open and Trap Closed

The transition from long-term synanthropic success to a synanthropic trap involves thresholds for key habitat conditions, for example, a threshold level of temperature or relative humidity during the life cycle, or the abundance or quality of a food resource. Changes in the habitat may not be uniform, and refuges in the environment that have suitable conditions may exist for a short time. Populations in refuges may be sustained by epigenomic adaptions. But synanthropic populations would disappear if there were insufficient time for genomic or epigenomic adaptions to new conditions. For

Fig. 8.1. Old house borer adult emerging from gallery in structural wood. Credit: W. Robinson

DOI: 10.1079/9781800626416.0008

Fig. 8.2. Old house borer larva in gallery in structural wood. Credit: W. Robinson.

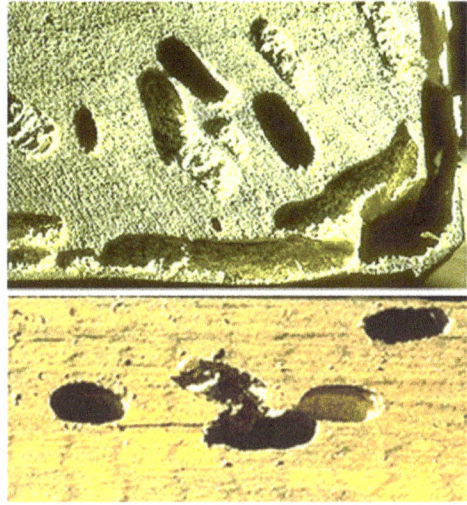

Fig. 8.3. Cross section of structural wood showing feeding galleries of old house borer larvae (top), and adult emergence holes in wood surface (bottom). Credit: W. Robinson.

any synanthrope, it would be difficult to predict the threshold for changes that affect survival of individuals or a population. A threshold would depend on which part of the habitat was the most vulnerable or has become unstable. A threshold might unfold in response to human-caused changes, which may be anticipated and more predictable than climate (temperature) changes.

Old House Borer

The old house borer infests wood (*Pinus* spp.) in built structures (Fig. 8.2). Larvae feed in wood with a 10–20% moisture content, and development is completed in 3–5 years. Dispersal in the urban biome is typically with new or used infested lumber; adult beetles live only a short time and have limited flight abilities. In European countries, this beetle primarily occurs in houses older than 10 years. In the USA, it typically infests houses less than 10 years old. These differences are linked to cultural and economic practices of reusing wood to build new houses. The common name, old house borer, is based on many household infestations originated from timbers taken from previously built (old) houses.

Pest status

The pest status is based on the loss of structural integrity of the infested wood pieces, and cosmetic aspects of infested wood and damaged

surfaces. Aesthetic aspects of an infestation include the adult emergence holes in exposed wood, and the audible sound of large (150 mg) larvae feeding in galleries close to the wood surface (Fig. 8.3). An infestation may be eliminated or end with the initial larval infestation, but this may require several years of feeding. Discovery of old emergence holes in attic or crawl-space locations may be misinterpreted as signs of active infestations. New emergence holes appear in spring when the moisture content of structural wood begins to increase.

The old house borer has been introduced into all major continents, and is a structural pest in eastern North America, England, Germany, Denmark, and South Africa. In these countries, it is capable of surviving in processed wood with a moisture content of about 25%. The biological potential of the old house borer provided the ability to infest wood with a moisture content unacceptable to other cerambycid beetles. This capability provided an abundant food source in primitive forest habitats. The food and the range of environmental conditions was the synanthropic basis for this species. It was successful until human-induced change eliminated the reservoir populations and the means of dispersal of infested wood to new habitats. The elimination of the food–habitat–dispersal triangle was a cultural change that was intolerable for the old house borer.

Climatic changes, along with the harvesting of natural stands of trees, eliminated the original forest habitats in north-western Africa and the Mediterranean region. Old house borer populations have disappeared from natural habitats in mainland Europe and Great Britain. It avoided extinction by transitioning from natural dead wood to seasoned lumber. In the USA, scrap pieces of lumber in small lumber-processing sites were often infested, and these populations served as reservoirs in the urban environment. In Europe, the house borer also used the artificial habitat of old houses, but it was occasionally encountered in forest habitats.

European infestations

Trap open

The presence of the old house borer in European countries began with the reuse of infested wood from existing (old) houses in new house construction or refurbishing. Continued presence was based on outdoor dispersal of adult beetles from infested to new locations with seasoned pine lumber. Along with the trend to convert open attics to living space, beetle infestations decreased. Improved chemical and non-chemical control measures and improved quality of structural lumber changed the conditions for the old house borer in mainland European countries and the UK. The synanthropic trap began to slowly close as the suitable conditions, availability of food, and reservoir populations decreased.

Trap closed

A strong decline in infestations in Europe in the last 80 years has been documented. New infestations in Germany were at the rate of 55,000 per year between 1910 and 1960 but declined to 6000 new infestations in 2015. In the UK, houses built between 1921 and 1950 had high infestations, but after 1961 infestations decreased. The decline is broadly linked to effective chemical and non-chemical control

methods, the quality of wood used for house roof construction, and the pre-construction design of attic space as, or its conversion into, living space. This set of changes, working separately but concurrently in the urban biome, significantly changed the habitat and the availability of the food source. This was a human-generated change to the living space that probably could not have been avoided by the old house borer making survival adjustments, whether genomic or epigenomic, to the availability and features of the food source.

US infestations

Trap open

Records of the old hose borer begin in the early 1800s. Infestations likely originated from furniture or other wood material, such as shipping and storage crates, from the UK and Europe. House construction and furniture manufacturing at this time in the USA was from natural-growth pine trees processed in small lumber mills. Eventually, populations were established in scrap wood piles at the mills. Lumber of various dimensions was infested while at the mill storage yard (Fig. 8.4). Persistence of the old house borer in the USA was based on the distribution of mills that provided suitable conditions year-round and a reliable form of dispersal in infested lumber.

Trap closed

Within about 10 years, the habitat of the old house borer in the USA changed. The changes were based on economics and not on individual features or conditions in the habitat. Small lumber mills closed as lumber manufacturing moved to regional operations and stored wood was protected. Mill sites were cleared, and scrap pieces of pine lumber were recycled (Fig. 8.5). Chemical control measures to treat infested pieces improved, and many infestations ended with the initial larvae in the wood. This was an economic change that provided the old house borer few options for adjustment.

Fig. 8.4. Construction lumber stored outside at lumber mill. Credit: W. Robinson.

Fig. 8.5. Pieces of scrap boards at lumber mill. Credit: W. Robinson.

Casemaking and Webbing Clothes Moth

In the early 1800s in Europe, the UK, and the USA, clothes moths commonly occurred in the household habitat. The two synanthropic species were: the casemaking clothes moth (*Tineola pellionella*) and the webbing clothes moth (*T. bisselliella*) (Fig. 8.6). Casemaking moth larvae create and remain in a silken tube or case while they feed and develop. They are restricted to feeding on keratin-based food, such as wool, and require high (75%) relative humidity for development. Webbing moth larvae create a silken enclosure on a substrate and generally remain in it while

they feed and develop. These larvae feed on both animal and non-animal food resources and tolerate low (20%) relative humidity conditions.

Both clothes moth species survive in a broad range of temperature and relative humidity conditions, and both consumed keratin-based food material. As modern society gradually improved living conditions and clothing fabric, conditions in the household habitat changed. The changes included widespread use of central heating, and the culture-driven change to synthetic fibers. These changes primarily affected the development of the immature stages of clothes moths, as these are the only stages that feed. By the end of the 1800s, houses had installed or were built with central heating. The new indoor environment had higher year-round temperatures and decreased humidity. These changes in conditions resulted in the pest status of casemaking moths declining. But the long-term presence of the webbing moth in the modern household is uncertain, considering the limited use of natural fiber for clothing, and new control methods.

Household infestations

Trap open

Although the larval food base and development conditions were restrictive, the casemaking

Fig. 8.6. Larval case of webbing clothes moth on fabric. Credit: USDA.

moth was a common household pest species in Europe for decades. Early conditions in rural and urban houses provided little uniformity in indoor temperature and relative humidity, and textiles and clothing were primarily animal based. The continued presence of webbing moths indoors is based on tolerance of low humidity, and ability for water conservation. The feeding shelter attached to the substrate reduces water loss in habitats that have low relative humidity. This species can utilize plant-based food, and populations have a broader base for survival.

Trap closed

The basic household habitat changed with the introduction of central heating, and this trapped the casemaking moth in a habitat that it could not fit. By the end of the 1800s, room-to-room heating was common in houses. This heating method provided a uniform temperature and relative humidity, including in clothing storage spaces. Rayon was introduced in the 1800s, and nylon was introduced in 1935. The increased use of synthetic or blended textiles in clothing and other fabrics also contributed to the synanthropic trap for the casemaking moth.

Firebrat and Silverfish

The common silverfish, the long-tailed silverfish, and the firebrat are facultative synanthropes in the urban biome. All three species are adapted to indoor temperature and humidity conditions and have metabolic rates suited to periods without feeding. Despite their common occurrence indoors, there are distinct differences between these species. The firebrat inhabits sites with high temperatures (32–41°C) and high relative humidity (76–85%). The common silverfish inhabits sites with moderate temperatures (22–27°C) and high relative humidity (75–97%). The long-tailed species is in sites with moderate temperatures (~24°C) and low (55%) relative humidity.

Firebrat

This species is well adapted to high temperatures and inhabits locations such as steam tunnels, around water heaters and similar sites. The adults and immatures stages are highly resistant to water loss. Their ability to take up water from a surrounding subsaturated (45% RH) atmosphere to maintain water balance enables survival in high-temperature and dry locations.

Harborages are dark voids with a narrow (4.5–6 mm) entrance. An aggregation pheromone orients conspecifics to the harborage. Development of juveniles through early instars is in the harborage on food that is provisioned by adults. eggs are often deposited in the harborage and optimal juvenile development conditions are at 35°C.

Common silverfish

The common silverfish (Fig. 8.7) inhabits various residential and commercial structures worldwide. It feeds on carbohydrate- and protein-based material, but pest status is limited, as infestations are small and foraging occurs at night. The success and distribution of this species was established in households without central heating and modern insulation. This habitat provided optimal temperature and humidity for life cycles lasting more than a year and long-term infestations. Improved building construction, central heating and cooling, and other modern features resulted in less suitable conditions for development. Habitat changes created a synanthropic trap for the common silverfish. The continued presence of this once successful species may be limited to fringe habitats in the urban biome. It occurs in natural habitats in central and southern Europe and the Mediterranean region, but in other parts of the world it is an obligate synanthrope.

Long-tailed silverfish

Within the last 20 years, the long-tailed silverfish has significantly increased its distribution and pest status in many regions, especially in European countries. This change is linked to improvements and modernization of household conditions resulting in reduced humidity and moderated and controlled temperatures. These conditions match the habitat preferences of the long-tailed silverfish, and less so the common silverfish. The synanthropic trap that closed on the common silverfish was followed by a species with fitness traits suitable for the modern household habitat. The culture-wide changes to

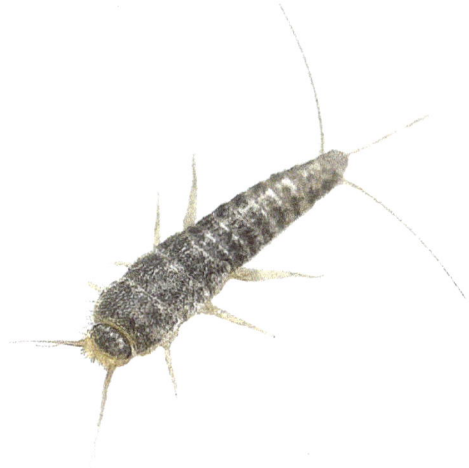

Fig. 8.7. Common silverfish. Credit: USDA.

household conditions influenced the distribution and persistence of these two species; one is in decline, and one is increasing.

Synanthropic Traps Now Closing

The concept of a synanthropic trap can be expanded from explaining what has already happened to a species to what is in the process of happening. The "open" trap positions can be examined and the "closed"' position predicted based on the expected habitat conditions. Beetles in a synanthropic trap that is slowly closing are species that feed on hardwood lumber used in flooring and beetles that feed on lumber used in house construction.

Powderpost beetles, *Lyctus* spp.

These beetles infest the wood used for hardwood flooring in residential and commercial buildings. Sections of flooring are infested in storage facilities before installation (Fig. 8.8). Storage facilities sustain populations of *Lyctus* powderpost beetles. In these storage sites, female beetles have available wood for egg laying. Pieces of flooring will become infested, and the adult emergence holes will appear after the flooring is installed in the house.

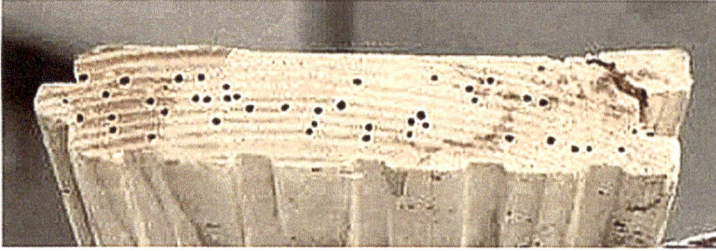

Fig. 8.8. *Lyctus* sp. powderpost beetle holes in flooring board. Credit: W. Robinson.

Trap closing

The demand for hardwood flooring has been declining in recent years in the USA and other countries. Composite wood technology produces materials that have the appearance of wood for less than the cost of genuine wood. The biggest change to hardwood flooring has been the invention of the engineered plank in the 1960s. Using a plywood or composite base, companies glue a veneer of the hardwood to the top, creating engineered hardwood, which is more stable than natural hardwood. This flooring trend is expected to continue, and the popularity of hardwood flooring and the populations of *Lyctus* powderpost beetles will decline. The synanthropic trap is slowly closing on *Lyctus* in the urban biome.

Powderpost beetle, *Euvrilletta peltata*

This is the most common beetle that infests structural wood. Females will lay eggs on, and larvae can feed on, hardwood and softwood lumber. House infestations begin when beetles fly to structural wood that is exposed in unfinished crawl spaces (Fig. 8.9). This space has year-round moderate temperature and high relative humidity because of the exposed soil beneath the flooring.

The wood moisture and crawl-space environmental conditions are ideal for infestations of *E. peltata*. Wood moisture content of 14–17% is ideal, but above 20% the conditions are suitable for wood-decay fungi and not beetle larvae. Wood moisture below 12% does not limit larval development after 3 months. Some of the lumber used for house construction comes from fast-grown pine trees which have a large amount of springwood, the preferred food of *E.*

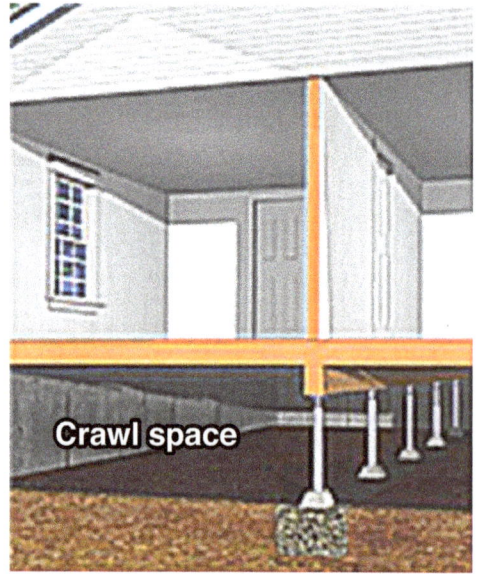

Fig. 8.9. Cross section view of house showing crawl space. Credit: W. Robinson.

peltata. These conditions existed for many years in houses with crawl spaces, and this beetle developed into a common pest.

Trap closing

Modern construction has resulted in a decrease in houses with crawl-space construction, and existing houses have undergone remedial repair that seals the soil surface. The result is a decline in suitable wood for this beetle. This type of house construction is limited to south-eastern USA, and in that region is declining. As that decline continues, the available wood suitable for *E. peltata* will decline. Eventually, crawl spaces in existing houses will be sealed and the trap closed for *E. peltata*.

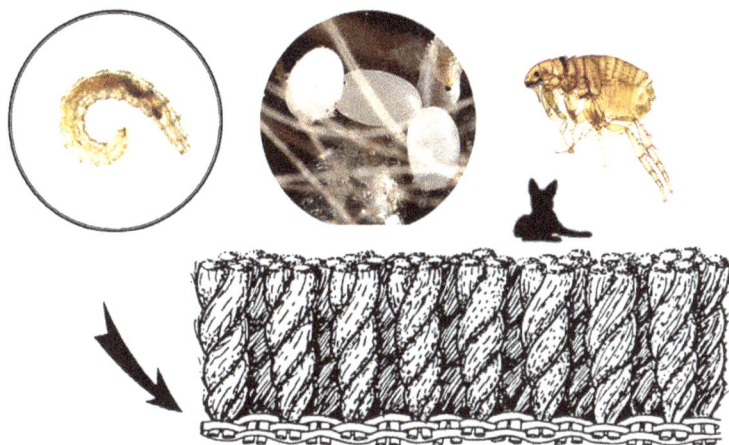

Fig. 8.10. Cross section of carpet showing adult flea, eggs, and larva. Credit: W. Robinson.

Cat flea, *Ctenocephalides felis*

This species is the primary blood-feeding pest of cats and dogs around the world. Success indoors is based on a harborage in which adults provide developing larvae a critical food resource. Pile carpeting is an optimal harborage for larvae because it provides ideal conditions for eggs, larvae, pupae, and flea feces (Fig. 8.10). It resists removal of fleas by vacuuming methods. While most of the eggs can be removed, they hatch within 48 hour and larvae move to the base of the pile where they are sheltered from vacuuing. Only 15% of the larvae can be removed by vacuuming.

Trap closing

Consumer demand for carpeting in modern housing is declining. Hard-surface flooring is replacing carpet in house construction and in floor renovation. The decline of and trend away from pile carpeting in houses in the USA and other countries may be linked to the amount of care required by pile carpeting and the potential for hosting arthropods and retaining allergens they produce. The preference of homeowners for non-carpet surfaces will influence the abundance of cat fleas indoors. Without this key harborage, the link between adult flea feces and successful larval infestations will weaken. The trap is closing on *C. felis*.

Varied carpet beetle and furniture carpet beetle, *Anthrenus verbasci* and *A. flavipes*

The larval stages of these two species infest and damage household furnishing and clothing with animal origin. Infestations are sustained by wool and blended clothing and blankets stored indoors. The high humidity, limited cleaning, and storage of wool products for long periods in early households enabled ongoing infestations. The optimal relative humidity for egg hatch and larval development is 70–80%. Adults can survive without feeding, but the overall environment, including relative humidity, can impact their longevity and reproductive success. Distribution of these beetles was primarily through the exchange and transfer of household materials to new locations.

Trap closing

The introduction of central heating provided uniform temperatures and lower humidity in houses and other structures. Consistently low humidity influences carpet beetle larvae by extending the development time and limiting growth of infestations. The decrease in use of animal-based clothing has limited the food available to beetle larvae. The growing preference for apartment living may limit the space available for storing clothing. These changes in household conditions are unfavorable for carpet beetles and are likely to eliminate the persistence of indoor populations of these two species.

Synanthropic Trap—Designed Disadvantage

The designed disadvantage concept is based on the creation of something to have a negative outcome. By applying this concept to ecological fitness, a synanthropic trap can be created by changing the conditions required for the survival and persistence of a species. A species that does not or cannot adjust to the change is trapped in a hostile habitat, and the populations are reduced or collapse. The successful use of a synanthropic trap depends on knowledge of the ecological fitness of a species to a habitat and resources, and the potential of making changes in these. The strategy for using this type of trap is to identify and change the links a species has to a habitat.

Harborage as resource

Harborage is an important feature for species occupying a habitat. Like food and water, harborage can determine the introduction and success of an invasive species, or the spread of an established species. This ecological and physical site is often overlooked when the fitness of a pest species to a habitat is considered. The assumption may be that a species can easily adapt to almost any available void space. But synanthropes may have narrow criteria when selecting a harborage, and this may be linked to their long-term success. A harborage site must be suitable for the expression of various aggregation and sex pheromones and volatile chemicals for harborage recognition used by some species. Management programs for pest species often include limiting or removal of food and some water resources (sanitation), or sealing so-called hiding places. But there is limited support data for some methods and limited success. Reducing eligible harborage can reduce the establishment of new populations and influence the current and next generation.

German cockroach populations are successful because they are based on a network of harborages in the infested habitat. Harborages are customized and conditioned by a group of conspecifics over an extended time. During this time, the offspring produced sustain the population in the habitat. The network of these productive harborages "anchor" a population by providing individuals through the addition of nymphs from eggcases. Adult German cockroaches have a relatively short lifespan and reproductive cycle for producing eggcases, which is indicative of a secure position in a habitat. While the harborages produce a genetically related group that may forage in a limited space, the number of these harborages defines the overall population.

Harborage advantage

The German cockroach harborage is a space with a preferred opening size that allows entry of nymphs, adults, and females carrying an eggcase. A narrow (3.2–6.2 mm) opening may reduce air movement inside and transfer of air with the outside (Fig. 8.11). This retains the high humidity conducive for development of small nymphs and reduces water loss during molting of all stages. Limited air exchange would retain and concentrate pheromones used to attract conspecifics or deter others from trying to enter the harborage. Pheromones in the feces have a unique scent based on the gut biome of the conspecifics. The pheromone is linked to the food eaten by individuals in the harborage. Crowding in the narrow space increases the temperature by about 0.6°C, and these conditions speed nymph development.

3.2–6.2 mm

Fig. 8.11. Cross section graphic of German cockroach harborage. Credit: W. Robinson.

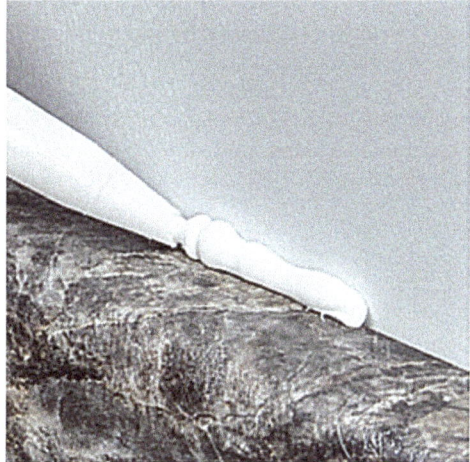

Fig. 8.13. Applying caulk to space between counter and wall. Credit: W. Robinson.

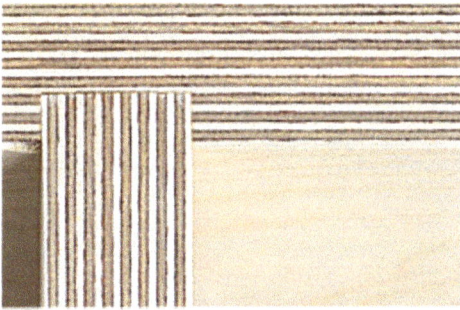

Fig. 8.12. Wood joinery in household cabinets. Credit: W. Robinson.

Harborage disadvantage

The initial and continued presence of German cockroach infestations in a habitat is based on the abundance of cracks and crevices that can be used as a harborage. Voids that have an opening of 3.2–6.2 mm and a depth of 11 mm, a rough surface, and are close to food and water are eligible. They can be conditioned with cockroach feces and pheromones to gradually fulfill the role of a breeding site. Preventing infestations would require limiting the voids that have the necessary dimensions and locations.

The objective of a designed disadvantage strategy would be to build cabinetry with 3.2–6.2 mm tolerances between structural pieces (Fig. 8.12). This design would reduce the voids and gaps eligible for harborage establishment. Designed disadvantage cabinetry installed in locations with an ongoing infestation would prevent establishment of new harborages and

influence the sustainability of the resident population. When installed in new locations, invasive German cockroach adults or nymphs would find a limited amount of suitable harborage. Without a secluded harborage, females would deposit eggcases in exposed sites and first-stage nymphs would have limited survival because of the lack of food. The number of eggcases produced and the eggs per eggcase decrease when no harborage is available. Introduced large nymphs and males have no orientation to a harborage after foraging.

The designed disadvantage strategy is directed to the harborage because this is the most important connection between the German cockroach and the habitat, and the most vulnerable. This feature can be manipulated on a wide scale and done once, such as changing all the cabinetry in a location. While the German cockroach has been capable of developing physiological resistance to modern insecticides, a complete change in behavior and conditions for development are not likely. Construction and manufacturing of cabinetry may continue to improve on limiting the spaces suitable for establishing a harborage. Without harborages to anchor populations of German cockroaches, infestations will be significantly reduced or absent.

Exposed cracks and crevices around sinks and cabinets are often misunderstood as functional harborages, and they are often sealed with a caulking compound (Fig. 8.13). But sealing or closing these exposed sites has little value as a control strategy. These narrow spaces may be used as temporary shelters by adults and large nymphs during foraging, but other spaces are usually available. Only about 41% of the internal narrow spaces inside cabinets are accessible for manual caulking, the remaining spaces are positioned where environmental conditions are suitable and where there are other eligible spaces for harborages. Similar caulking strategies are often advocated for preventing American and oriental cockroaches from entering from outside of buildings. There are numerous access points for these species, and caulking has not been a successful means of prevention.

Additional Reading

Auer, J., Opitz, C. and Kassel, A. (2022) Biological control of wood destroying beetles with *Spathius exarator* (Hymenoptera: Braconidae). In: Bueno-Marí, R., Montalvo, T. and Robinson, W.H. (eds) *Proceedings of the Tenth International Conference on Urban Pests*, ICUP, Barcelona, Spain, pp. 70–75. Available at: https://icup.org.uk/media/vm3jwgkz/8-auer-176-f-pp-70-75.pdf (accessed 29 September 2025).

Biebl, S. and Auer, J. (2017) Practical use of braconid wasps for control of the common furniture beetle (Coleoptera: Anobiidae). In: Davies, M.P., Pfeiffer, C. and Robinson, W.H. (eds) *Proceedings of the Ninth International Conference on Urban Pests*, ICUP, Birmingham, UK, pp. 367–375. Available at: https://icup.org.uk/media/izvfoktx/icup1232.pdf (accessed 29 September 2025).

Lea, R.G. (1996) Cockroaches in the UK: Designing buildings to reduce the risk of infestation. In: Wildey, K.B. (ed.) *Proceedings of the Second International Conference on Urban Pests*, ICUP, Edinburgh, UK, pp. 573–578. Available at: https://icup.org.uk/media/dajnuw4q/icup821.pdf (accessed 29 September 2025).

Murphy, R.G. and Todd, S. (1993) Towards pest free dwellings in the urban environment. In: Wildey, K.B. and Robinson, W.H. (eds) *Proceedings of the First International Conference on Urban Pests*, ICUP, Cambridge, UK, pp. 423–432. Available at: https://icup.org.uk/media/deaf52rm/icup658.pdf (accessed 29 September 2025).

Plarre, R., Busweiler, S., Haustein, V., Laar, C. and Haustein, T. (2022) *Korynetes caeruleus* (Coleoptera: Cleridae) for biological control of *Anobium punctatum* (Coleoptera, Ptinidae). In: Bueno-Marí, R., Montalvo, T. and Robinson, W.H. (eds) *Proceedings of the Tenth International Conference on Urban Pests*, ICUP, Barcelona, Spain, pp. 34–44. Available at: https://icup.org.uk/media/lyipplv5/6-plare-1-f-pp-34-44.pdf (accessed 29 September 2025).

Pospischil, R. (2017) *Lyctus* (Coleoptera: Bostrychidae): A never ending story. In: Davies, M.P., Pfeiffer, C. and Robinson, W.H. (eds) *Proceedings of the Ninth International Conference on Urban Pests*, ICUP, Birmingham, UK, pp. 377–380. Available at: https://icup.org.uk/media/w4ieq32f/icup1233.pdf (accessed 29 September 2025).

Robinson, W.H. (2002) Role of reservoir habitats and populations in the urban environment. In: Jones, S., Zhai, J. and Robinson, W. (eds) *Proceedings of the Fourth International Conference on Urban Pests*, ICUP, Charleston, USA, pp. 217–223. Available at: https://icup.org.uk/media/aswjw2cb/icup223.pdf (accessed 29 September 2025).

Rust, M.K. and Hemsarth, W.L.H. (2017) IGRs for cat flea (Siphonaptera: Pulicidae) control revisited. In: Davies, M.P., Pfeiffer, C. and Robinson, W.H. (eds) *Proceedings of the Ninth International Conference on Urban Pests*, ICUP, Birmingham, UK, pp. 253–258. Available at: https://icup.org.uk/media/j3jhmey5/icup1215.pdf (accessed 29 September 2025).

9 Pest Status and Climate Change

Climate change is anticipated to cause range expansion in some insect pests, but for others it may be increased population density, or decreased development time, which may result in increased generations in the habitat. This global condition has already or soon will alter the established range limits of disease vectors and their host animals. The ability to adjust to new conditions, whether linked to climate or humans changing the habitat, drives the spread of successful species in the urban biome. Adaptations to change can be physiological, developmental, and include behavioral traits. For many of the important structural pests or disease vectors, such as mosquitoes and ticks, there is historical data on their numbers and distribution data. The value of previous data is that it enables the ability to currently report range expansion and contraction over a wide geographic area for these species.

The basis of pest status for a species includes long-term persistence in the habitat, a real or perceived medical threat, economic loss, or simple aesthetic dislike in the living space. The continued presence of pests in the urban environment is due, in part, to the relative ineffectiveness of control or management measures, and the existence of reservoir populations that enable reinfestation. Long-term persistence of most arthropod pests is based on a network of these reservoirs and satellite infestations in relatively stable habitats. Pest management programs usually target the accessible infestations but not the reservoirs.

Climate warming affects the physical environment, such as the temperature of soil and standing water, regional or local extreme rain events, and, in turn, the habitats and associated animals (Fig. 9.1). These changes in conditions could profoundly affect the population dynamics

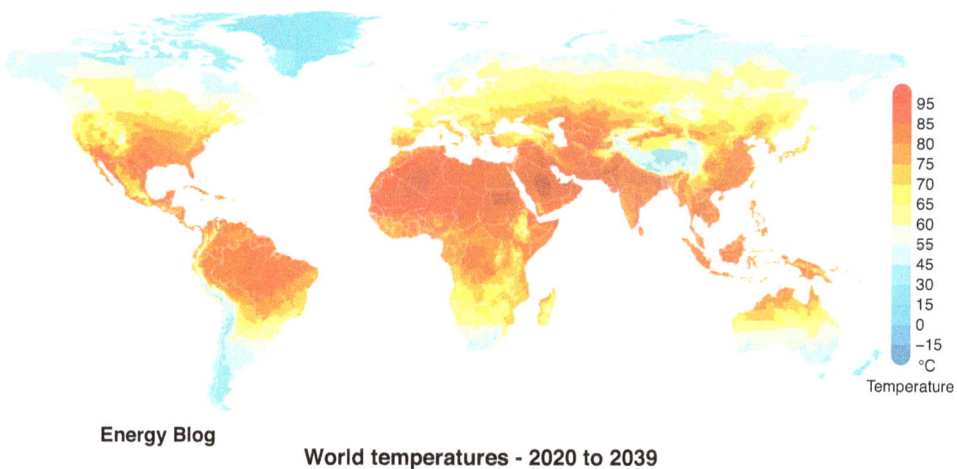

Energy Blog

World temperatures - 2020 to 2039

Fig. 9.1. World temperature projections for 2020 through 2039. Credit: World Meteorological Organization.

DOI: 10.1079/9781800626416.0009

of arthropods and other animals over a wide area. Anticipated levels of climate change could cause shifts in population density and seasonal abundance, and increased generation time. Even short-term change could cause unpredictable pest outbreaks and quickly shift the geographical ranges of species. Climate change may impact pests in the urban biome by creating or expanding habitat sites for species whose habitat was limited by previous climate conditions.

Climate Change and Urban Rat Populations

An increase in the pest status of the brown rat is expected with climate change. Warming temperatures coupled with the urban heat island effect may extend the foraging and breeding period for rats. Increasing urbanization and population size of cities will provide more food waste, which supports increases in brown rat populations (Fig. 9.2).

The human population in cities is projected to increase by 25% by 2050. Data from 16 global cities confirmed that higher human population density is linked to a growth in rat numbers. Warmer temperatures and lower winter mortality provide rats longer periods of foraging. Cities with large and increasing human populations have a trend of increasing rat sightings. Extensive rodent control programs in some large cities may be reducing rat sightings and show an overall negative trend.

Cities that lost vegetation between 1992 and 2020 experienced large increases in rats. This may be due to an increase in new urban residences, food service establishments, and the refuse and food waste generated as a result. Vegetation loss reduces shading and increases heat retention, which enhances the urban heat island effect. The increase in rat populations may be linked to both the habitat preferences of rats and food availability. Brown rats prefer access to bare ground for their burrowing, and large green spaces may have less bare ground. Large spaces may have less food waste available, or it is centered. Brown rats can use patches of bare soil or nest within discarded boxes close to restaurant garbage bins. Strategies to slow the increase of rat numbers include managing food waste, using rodent-proof dumpsters, frequent garbage collections, and food-waste diversion programs.

Temperature and Disease Vectors

There is a growing base of information on the effect of climate warming on insects. The influence of temperature increase on insect populations is gradual and only evident after distribution data becomes available. Changes can directly affect growth, reproduction, and survival within a thermal range. Indirect effects include the abundance of vertebrate or plant hosts. For some species, it will be faster development and a potentially increased number of generations per year. When this includes mosquitoes and ticks, their medical pest status will increase. Outdoor species such as the outdoor populations of American and oriental cockroaches may experience accelerated growth and development, but food resources will have to match these changes.

Mosquitoes

It is generally expected that climate change, particularly increases in temperature and changes in rainfall patterns, will affect the distribution of mosquito vectors and the spread of the diseases they transmit. The projected changes in the distribution of the yellow fever mosquito, *Aedes aegypti*, provide an example of the outcome of global warming for this important vector.

Yellow fever mosquito

This mosquito is primarily found in tropical and subtropical urban areas. The range expands and contracts seasonally in the USA along with the temperature-limited survival conditions. Increasing temperature and precipitation, and the prevalence of trash in urban environments, affect the survival and distribution of *Ae. aegypti*. This synanthropic species is successful in breeding in water-filled containers and trash that holds water. Warmer temperatures enable faster

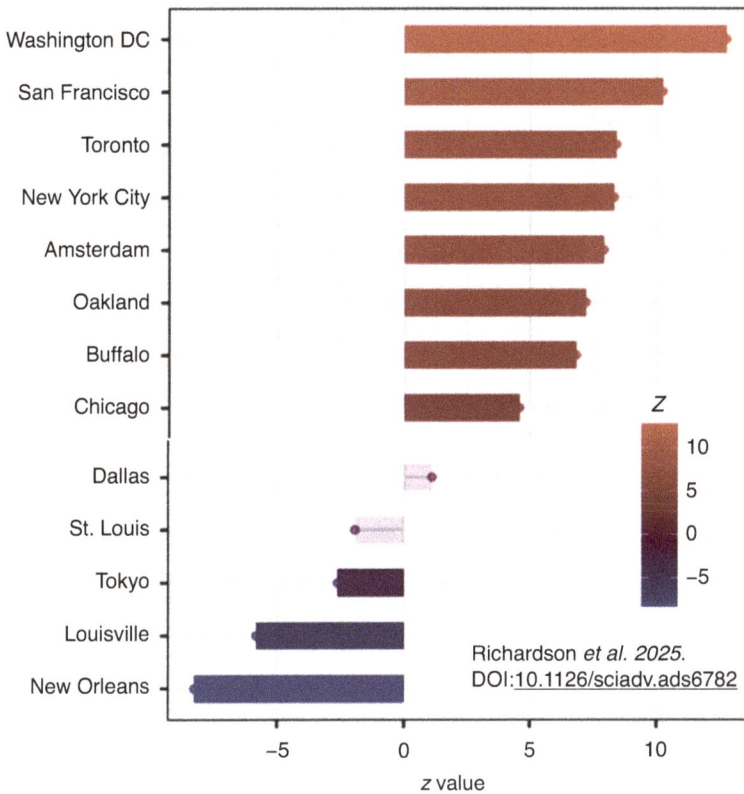

Fig. 9.2. Trends in rat sightings across 13 cities with long-term data on rat complaints and municipal inspections. Positive *z* values represent increasing rat numbers, negative values indicate decreasing trends. Credit: Richardson *et al*. 2025.

development of the larval stages, and greater survival of the adults. Increased precipitation provides water needed to complete the aquatic larval stages. The range of *Ae. aegypti* is expected to expand, based on temperature increases and the availability of urban breeding sites (Fig. 9.3). Urban populations continue to increase, and the breeding habitats influenced by urban heat islands will continue to benefit this species.

Much of the USA and parts of southern Canada are projected to be suitable for the yellow fever mosquito by 2100. The ecological range of this mosquito is projected to continue northward following climate change and increased precipitation in the region. The current and simulated (future) ecological niche for this mosquito is influenced by increased average temperature in January. Temperature is linked egg-to-adult

survival and persistence of local populations. Low winter temperatures prevent surviving the season and limit self-sustaining local populations. None of the life stages survive below 0°C. Once the average minimum daily temperature is above 0°C, the suitable habitats for this mosquito increase.

The impact of global warming on *Ae. aegypti* (Fig. 9.4) distribution may be at the geographical limits of current distributions, where conditions limit suitable and consistent breeding sites. Records of *Ae. aegypti* in northern locations where conditions are suboptimal may be from populations that require reintroduction during summer. Increasing temperatures associated with climate change will likely create new suitable habitats at the edge of the current range.

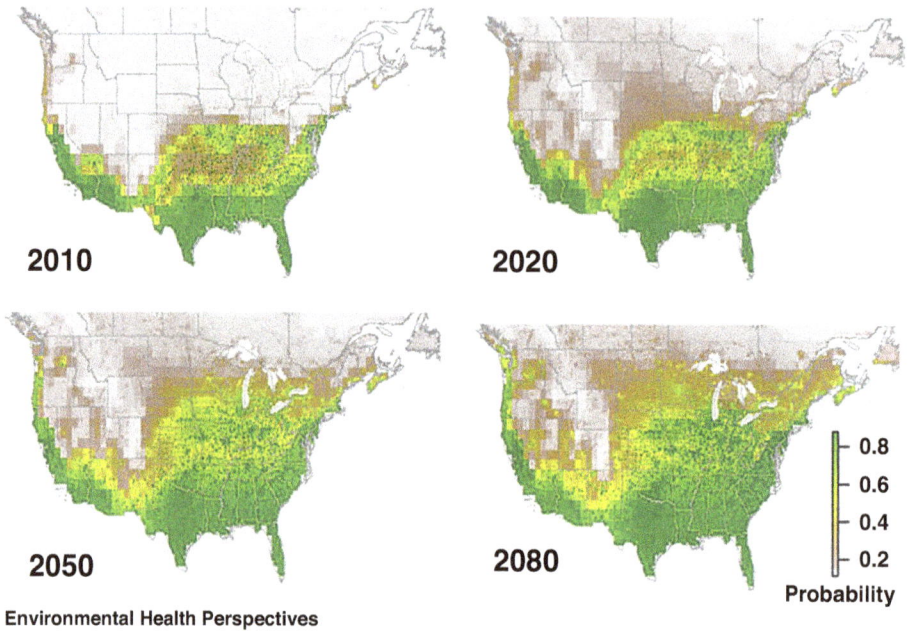

Environmental Health Perspectives

Predictions of *Aedes aegypti* range expansion based on regional climate model data sets from years 2006 to 2100.

Fig. 9.3. Predictions of *Aedes aegypti* range expansion based on regional climate model data for years 2010 to 2080. Credit: Environmental Health Perspectives.

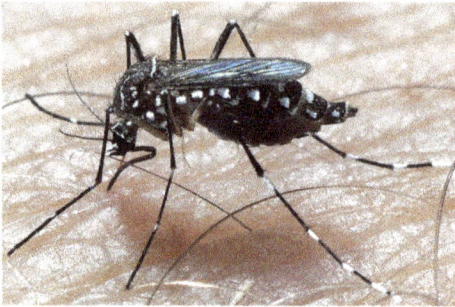

Fig. 9.4. *Aedes aegypti* mosquito. Credit: CDC US.

Ticks

Tick biology and distribution are influenced by climate temperature and humidity. Global warming might benefit some tick species that are adapted to humid environments but probably will have a minor impact on most other ticks. The most noticeable impact of temperature change is the influence on the geographic distribution and abundance of insects. For example, the blacklegged tick (Fig. 9.5) has spread from north-eastern USA into south-eastern Canada. This northward range expansion coincided with climate warming in Canada, which was about double the magnitude of global warming. It included extreme heat events, less extreme cold, and shorter periods of snow and ice cover. The geographic distribution of this tick in the USA has expanded and has more than doubled over the past two decades.

American dog tick

The American dog tick (Fig. 9.6) is the primary vector for the bacterium causing Rocky

Fig. 9.5. Blacklegged deer tick adult. Credit: Pixabay/CC0 Public Domain.

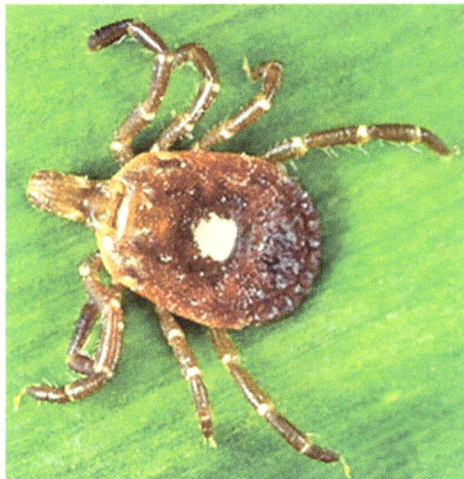

Fig. 9.7. Lone star tick. Credit: CDC US.

Fig. 9.6. American dog tick adult. Credit: Pixabay/CC0 Public Domain.

Mountain spotted fever in humans and domestic animals. The immature stages feed on small mammals; the adult feeds on larger mammals, including humans and household pets. The host-questing behavior of this tick is to climb to the tips of vegetation along woodland pathways and trails. The common hosts in urban green spaces are white-footed mice and deer mice. These host animals are successful in urban green spaces by utilizing discarded human food and habitats in leaf litter and ground vegetation. The American dog tick has limited overwinter

survival in regions where mean winter temperatures remain below 0°C. Climate change has resulted in milder winter temperatures, with almost all the north-eastern USA and much of the mid-western region averaging above freezing temperatures during winter months.

Lone star tick

The lone star tick (Fig. 9.7) can transmit to humans a bacterium that introduces the alpha-gal molecule that causes alpha-gal syndrome, which is a food allergy to red meat. Between 2010 and 2022, according to the US Centers for Disease Control and Prevention, there were more than 110,000 suspected cases of alpha-gal syndrome in the USA. This condition has been reported in European countries, and the primary tick vector is thought to be the European deer tick. The lone star tick will actively seek a host that it has detected several meters away. This behavior of recognizing host odors enlarges its feeding range because it is not waiting on vegetation for a host to pass by. Reports show that the lone star is expanding its zoogeographic range in northern and mid-western USA (Fig. 9.8). This tick is sensitive to changes in microclimatic conditions, especially ground-level moisture, which is linked to atmospheric relative humidity. Arid and semi-arid habitats in the USA will likely limit

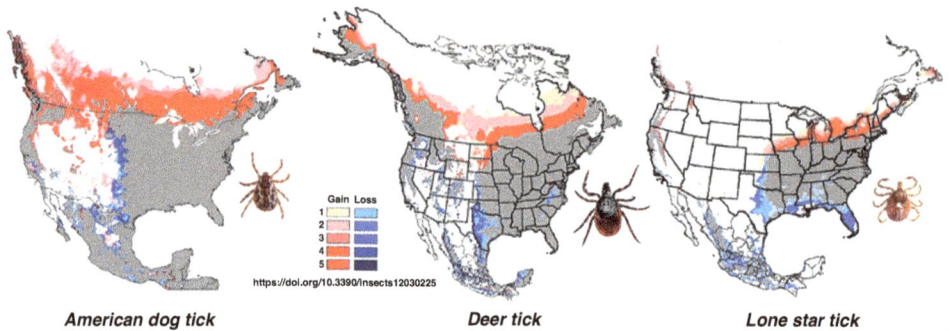

Fig. 9.8. Predicted US range expansion of American dog tick, deer tick, and lone star tick. Credit: Alkhise *et al.* 2021.

its westward range expansion. Climate change will likely cause dry summers in some regions and influence the range of this species.

Blacklegged tick

This is the primary vector for more tick-borne diseases than any other tick in North America. It transmits the bacteria causing Lyme disease and tick-borne relapsing fever and the virus causing Powassan illness. The geographic range of this tick has expanded due to warmer winter temperatures. It has spread throughout most of eastern USA and matches the distribution and abundance of white-tail deer. The number of populations have also intensified in many of the colonized localities. Males and females are active when the daytime temperature remains above freezing. Adults prefer questing on the tips of leaves on low-growing shrubs. Fully engorged females drop off the host into the leaf litter, where they can overwinter. Females lay up to 2000 eggs in a single mass. Larvae emerge in summer (Fig. 9.9).

European deer tick

The European deer tick (Fig. 9.10) is the primary vector for tick-borne diseases throughout Europe. The geographical and ecological distribution has changed in recent years, and it now occurs in habitats at high altitude and

latitudes. It transmits several diseases, including the bacterium that causes Lyme disease, human granulocytic anaplasmosis, tularemia, babesiosis, and Tribec virus. It feeds on a range of mammals, birds, and reptiles and frequently bites humans. The life cycle is completed within 3 years but is shorter with optimal conditions and when suitable hosts are available. This small (2.4–3.6 mm) tick prefers habitats with high (80%) humidity, which includes sites in urban parks and woodlands. After mating, a female drops off a host to the ground and remains in the area for the 8 weeks needed for egg production. Up to 2000 eggs are deposited at the same time, and larvae hatch in about 8 weeks. Larvae do not move horizontally over a large distance and remain aggregated around the hatch site. Seasonal abundance reaches maximum density in spring or autumn.

Brown dog tick

The brown dog tick (Fig. 9.11) is distributed around the world. It is unusual among ticks because it can complete its entire life cycle in habitats indoors or outdoors. Infestations develop and persist in households that have companion animals. Although it will feed on a variety of mammals, dogs are the preferred host. Each active stage feeds only once, then leaves the host to digest the blood meal and molt to the next stage. Indoor infestations enable feeding to occur on the same host animal for the

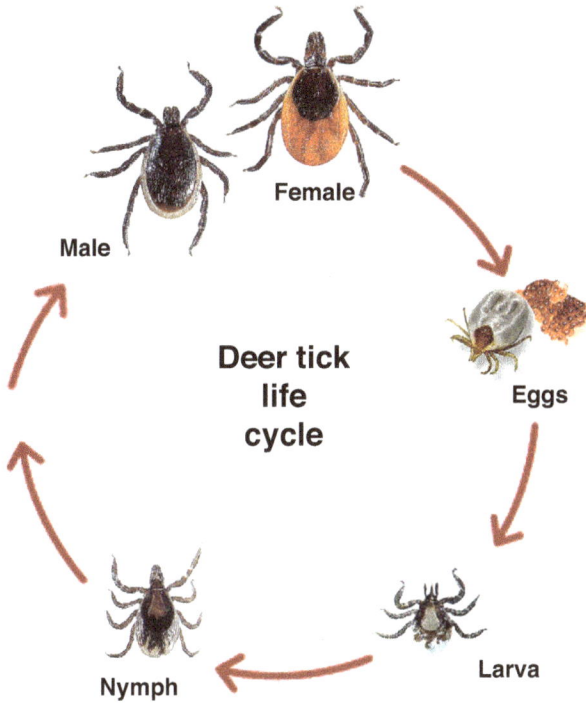

Fig. 9.9. Deer tick life cycle. Credit: W. Robinson.

Fig. 9.10. European deer tick. Credit: Pixabay/ CC0 Public Domain.

Fig. 9.11. Brown dog tick adult. Credit: Pixabay/ CC0 Public Domain.

entire life cycle. A fully fed female can lay over 4000 eggs, and cracks and crevices in houses, garages, and dog runs provide egg-laying sites. First-stage larvae hatch in about 12 days and begin to search for a host; they feed for about 10 days then drop from the dog. Nymphs attach to the next host (possibly the same dog), feed for about 13 days, drop from the dog, and then develop into adults. Males and females attach to hosts and feed, although males feed only for short periods (Fig. 9.12).

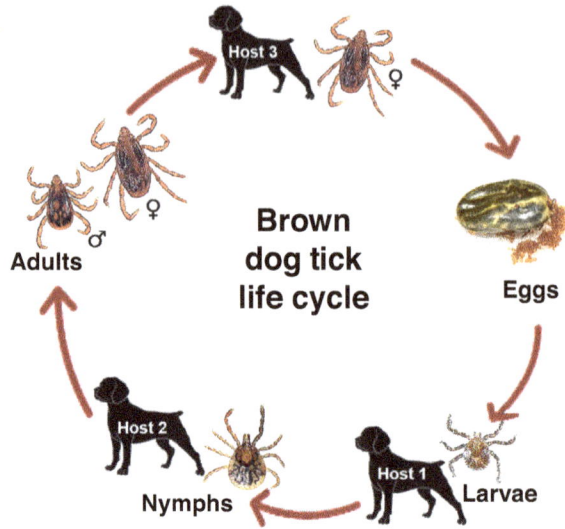

Fig. 9.12. Life cycle of brown dog tick. Credit: W. Robinson.

In outdoor habitats, such as urban green spaces, this tick can feed and survive on small mammals during warm months but is often unable to overwinter there. The brown dog tick is adapted to warm and dry environments. It is sustained by patches of vegetation in urban parks where daytime temperatures are moderated by vegetation and humidity is usually high. Urban residents may seek relief from high temperatures in urban parks and may take their pets with them. Walking in vegetation off paths and walkways can bring people and pets to sites occupied by ticks searching for a blood meal. The brown dog tick prefers dogs, and the availability of these animals in urban parks not only supports populations there but may spread infestations to new (dog) hosts.

There is an increased risk of humans being bitten by ticks as the environmental temperatures increase. In the temperate populations of the brown dog tick, high temperatures increase the preference for attaching to humans rather than dogs. This suggests that changing climate conditions may increase human bites and disease transmission (Fig. 9.13). Hot weather events may be expected to increase cases of disease transmitted by the brown dog tick. The risk of disease further increases since the brown dog tick is capable of living indoors and

this species is widely spread around the world. Rickettsial disease and reports of increased aggression during heat waves in Europe may be due to an increase in ticks preferring to feed on humans. The annual number of days with temperatures over 38°C is expected to increase across the continental USA in the next decade.

Urban Parks and Ponds

Cities and large metropolitan areas usually include planned natural recreation areas such as parks and ponds (Fig. 9.14). These spaces may be fragments of the original landscape with natural vegetation, ground cover, drainage, and associated animals. Other sites may be built to provide the same features and located to provide public access. The benefit of green and water-based (blue) spaces is wildlife biodiversity and the mental health benefits and wellbeing provided by experiencing natural habitats and animals. Depression, anxiety, obesity, and heatstroke are more prevalent in urban areas that lack access to shady tree canopies and green open spaces. Interaction with green space may include individuals exploring away from the paved walkways or

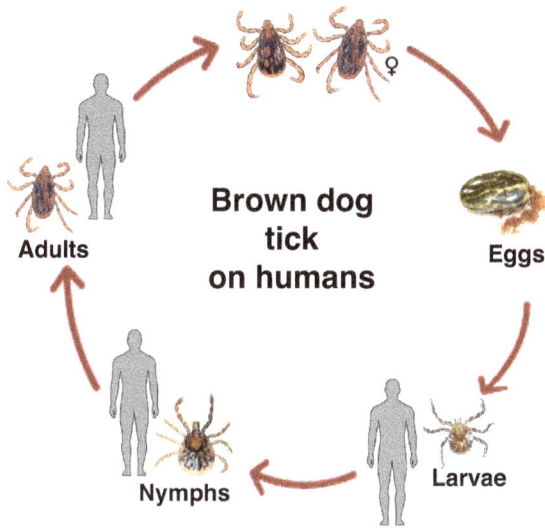

Fig. 9.13. Life cycle of brown dog tick on humans. Credit: W. Robinson.

Parks and ponds

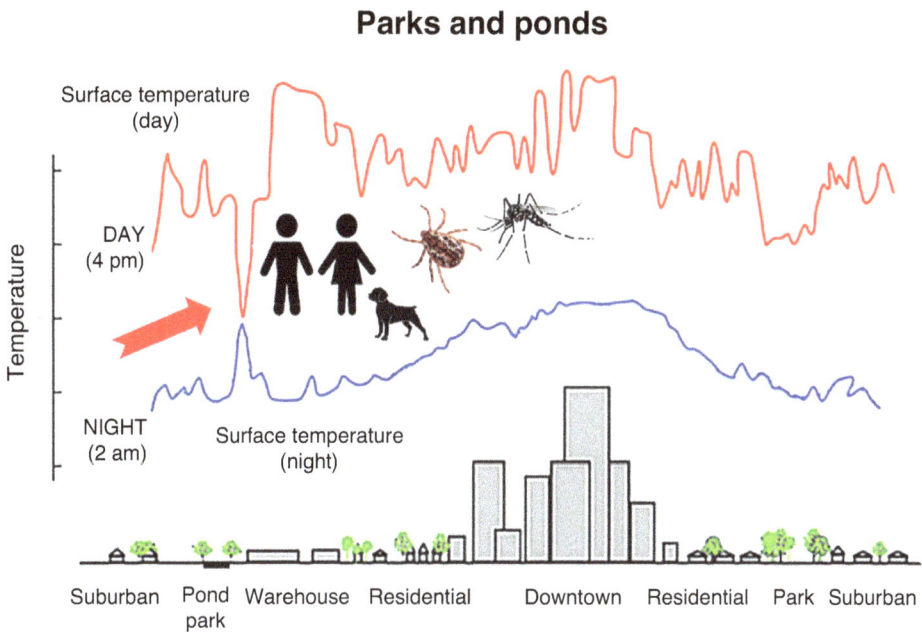

Fig. 9.14. Temperature range in urban parks and around ponds. Credit: Epa.gov.

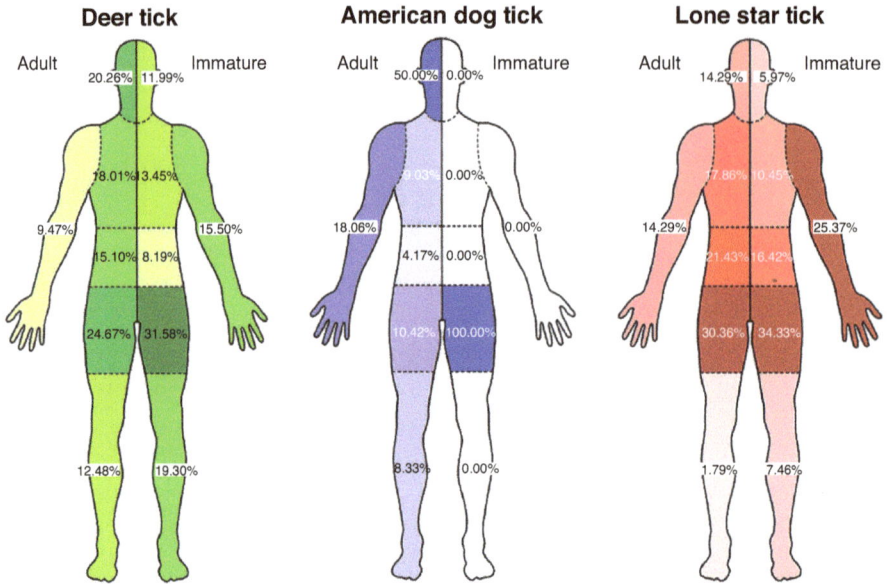

Fig. 9.15. Distribution of tick species attachments on the human body. Credit: Hart et al. 2022.

Fig. 9.16. Tick biting into skin. Credit: W. Robinson.

Ticks

Ticks are ectoparasites of terrestrial vertebrates and colonize natural habitats where temperature and humidity are suitable for long-term populations. They are also successful in the urban environment when those conditions are available for them and there are humans in the same location. Ticks are common in recreation areas and linked to the spread of disease in these sites. Deer ticks transmit Lyme disease, tularemia, babesiosis, and Tribec virus. Adults and nymph stages of the American dog tick and brown dog tick can attach to dogs in urban parks, and the lone star tick can attach to humans (Fig. 9.15).

The most likely human exposure is the bite of a single infected tick, whether adult or nymph (Fig. 9.16). The transmission of the disease agents increases with the length of time the tick remains attached to a host. For Lyme disease, the transmission time is 24–72 hours after the tick is attached. However, it is possible that transmission of the Lyme disease organisms could occur within 24 hours. The bites of deer ticks often remain

walking with dogs and cats. This creates the potential for coming into contact with mosquitoes and ticks established in the habitat. The populations of songbirds and small mammals in urban parks provide hosts for mosquitoes and Ixodid ticks.

undetected because of the small size of these ticks, especially the nymphs. This further increases the vector potential of this tick, as urban residents may have little experience in searching for and removing ticks.

Mosquitoes

To reduce the urban heat island effect and manage stormwater runoff, some cities are establishing wetland areas and expanding parks and ponds as long-term strategies to adapt to severe climate events. Some of these strategies may affect populations of mosquitoes and be detrimental to park visitors and residents in adjacent residential buildings. The house mosquito, *Culex pipiens* (shown), is one of the most common mosquitoes in urban and suburban habitats (Fig. 9.17). It is successful in urban areas because it utilizes the abundance of discarded containers around buildings, rainfall gutters, and stagnant water in street drains for breeding sites. This evening- and nighttime-biting mosquito is the primary vector for West Nile virus (WNV) around the world.

Females typically take blood meals from bird species, including American robins, European starlings, crows, and pigeons. Most of the feeding is at night while the birds are roosting. Urban crows are a competent host for WNV as they are easily infected and amplify the virus while it is in their body. Pigeons are a significant part of the *Cx. pipiens* diet, but a pigeon as a host is based on its availability within the 0.4–2.4 km of the female mosquito's flight range. In some cities, adult *Cx. pipiens* are more abundant in park sites, while the gravid females and larval stages are more abundant in residential areas. Bird hosts resting in trees and available for a blood meal may be abundant in parks and other green spaces, while stagnant water in artificial containers may be more abundant in residential areas. Dense vegetation in parks modifies temperature, humidity, and wind speed, which are all linked to adult mosquito survival.

Residential areas generally have stagnant water in artificial containers, trash scattered around buildings, and clogged street drains (Fig. 9.18). All of these are suitable larval breeding sites for *Cx. pipiens*, especially after severe

rain events. Female mosquitoes deposit eggs in any potential breeding site in residential areas, but some of these are short-term in seasonal urban heat island conditions. The large size of residential zones and variety of discarded containers that collect rainwater may attract gravid females in parks searching for egg deposition sites. Blood-fed females can fly more than twice as far as unfed females and are not limited to sites close to parks.

Increasing green spaces and enlarging parks and ponds adjacent to residential areas to reduce the urban heat island effect may not increase the risk of exposure to *Cx. pipiens* mosquitoes. The adult portion of a population may remain stable. The residential portion of a population, which is the larval stages, would be vulnerable to programs that reduced urban litter that provided short-term breeding sites. Eliminating some of these breeding sites would have a direct impact on limiting the spread of WNV in urban areas.

Red imported fire ant

Climate change can affect the spread of invasive species, but it will not be the primary driver. For invasive insects to be successful, they must adapt to the new environment, reproduce, and establish ongoing populations. The effects of climate change on critical features, such as dispersal and survival on food resources, are complex and can be positive or negative. The tolerance of new conditions, such as overwinter temperatures, by invasive insects is often greater than for native insects. This can facilitate their expansion to

Fig. 9.17. *Culex pipiens* female. Credit: Pixabay/ CC0 Public Domain.

Fig. 9.18. Storm sewer drain on street holding water. Credit: Pixabay/CC0 Public Domain.

Fig. 9.19. Red imported fire ant worker. Credit: USDA.

new habitats. It is uncertain how insect pests, both native and invasive, will respond to global warming, and it cannot be guaranteed that the warmer temperatures will be advantageous.

The red imported fire ant (RIFA) (Fig. 9.19) was originally from South America and accidentally introduced to the USA in the 1930s. Since then, it has spread to infest more than 367 million acres (148.5 ha). It is a major ecological and socio-economic problem that causes an estimated US$6 billion in economic damage annually. Negative impacts include a painful sting to humans and animals, damage to crop seeds, and negatively affecting the survival of native ants. In its original or introduced range, this ant favors disturbed sites, including public lands and urban and suburban parks. The occurrence of RIFA in the USA varies from year to year and is based on changes in temperature, moisture, and land management practices.

Temperature and precipitation are the main constraints on RIFA spread. After it was introduced and became established in the USA, it was expected that populations would not expand beyond south-eastern USA. This reasoning was based on adaptation of RIFA to their native subtropical climate, which would make them unable to survive cold weather. Winter temperatures with a minimum below $-3.7°C$ limit colonies' ability to produce adults because freezing reduces reproductive output. Dry conditions and cold temperatures in the north have slowed the range of its spread, but it is predicted to continue to expand (Fig. 9.20).

Current

2050

2070

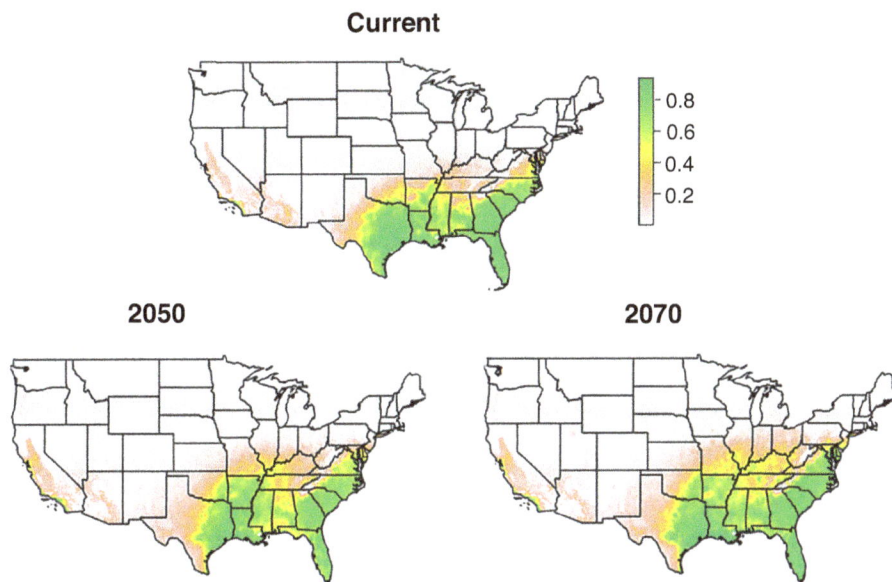

Fig. 9.20. Range expansion from current to 2050 and 2070 predicted for red imported fire ant in USA. Credit: Malone *et al.* 2023.

Additional Reading

Basseri, H.R., Yosafi, S. and Moosakazemi, S. (2005) Blood preferences of malaria vectors in Tehran. In: Lee, C.-Y. and Robinson, W.H. (eds) *Proceedings of the Fifth International Conference on Urban Pests*, ICUP, Singapore, pp. 275–279. Available at: https://icup.org.uk/media/b3ap4xdb/icup043.pdf (accessed 29 September 2025).

Dhang, P. (2017) Review of climate change impacts on urban pests. In: Davies, M.P., Pfeiffer, C. and Robinson, W.H. (eds) *Proceedings of the Ninth International Conference on Urban Pests*, ICUP, Birmingham, UK, pp. 25–31. Available at: https://icup.org.uk/media/w1kgal0a/icup1179.pdf (accessed 29 September 2025).

Merchant, V. (2022) Sustainability of IPM strategies in the current environment of pest resurgence. In: Bueno-Marí, R., Montalvo, T. and Robinson, W.H. (eds) *Proceedings of the Tenth International Conference on Urban Pests*, ICUP, Barcelona, Spain, p. 388. Available at: https://icup.org.uk/media/2eokudb1/4-merchant-72-pg-388.pdf (accessed 29 September 2025).

Mueller, G., Luescher, I.L. and Schmidt, M. (2008) Temporal changes in the incidence of household arthropod pests in Zurich, Switzerland. In: Robinson, W.H. and Bajomi, D. (eds) *Proceedings of the Sixth International Conference on Urban Pests*, ICUP, Budapest, Hungary, pp. 15–21. Available at: https://icup.org.uk/media/y1ub0jib/icup912.pdf (accessed 29 September 2025).

Neoh, K.-B. and Bong, L.-J. (2017) Factors driving *Paederus* outbreaks in human settings: Climatic factor or human intervention? In: Davies, M.P., Pfeiffer, C. and Robinson, W.H. (eds) *Proceedings of the Ninth International Conference on Urban Pests*, ICUP, Birmingham, UK, pp. 199–202. Available at: https://icup.org.uk/media/n0uinnbc/icup1206.pdf (accessed 29 September 2025).

Uspensky, I. (2008) Ticks (Acari: Ixodoidea) as urban pests and vectors with special emphasis on ticks outside their geographical range. In: Robinson, W.H. and Bajomi, D. (eds) *Proceedings of the Sixth International Conference on Urban Pests*, ICUP, Budapest, Hungary, pp. 333–347. Available at: https://icup.org.uk/media/kjynudbl/icup893.pdf (accessed 29 September 2025).

Uspensky, I. (2014) Conditions of tick (Acari: Ixodoidea) population persistence in the urban environ-
    ment. In: Müller, G., Pospischil, R. and Robinson, W.H. (eds) *Proceedings of the Eighth International
    Conference on Urban Pests*, ICUP, Zurich, Switzerland, pp. 203–210. Available at: https://icup.org.
    uk/media/lc4d3tnd/icup1112.pdf (accessed 29 September 2025).
Van Wijnen, J.H., Verhoeff, A.P. and Otten, G. (1996) Cockroach allergen in house dust and specific IgE
    to cockroaches. In: Wildey, K.B. (ed.) *Proceedings of the Second International Conference on Urban
    Pests*, ICUP, Edinburgh, UK, pp. 231–234. Available at: https://icup.org.uk/media/yuhl44v3/icup730
    .pdf (accessed 29 September 2025).

Bibliography

Chapters 1, 2, 3

Aronson, M.F., Lepczyk, C.A., Evans, K.L., Goddard, M.A., Lerman, S.B. *et al.* (2017) Biodiversity in the city: Key challenges for urban green space management. *Frontiers in Ecology and the Environment* 15, 189–196. DOI: 10.1002/fee.1480.

Chu, E.K., Fry, M.M., Chakraborty, J., Cheong, S.-M., Clavin, C. *et al.* (2023) Built environment, urban systems, and cities. In: Crimmins, A.R., Avery, C.W., Easterling, D.R., Kunkel, K.E. and Stewart, B.C. (eds) *Fifth National Climate Assessment*. U.S. Global Change Research Program, Washington, DC. DOI: 10.7930/NCA5.2023.CH12.

Clements, F.E. and Shelford, V.E. (1939) *Bioecology*. J. Wiley & Sons, New York.

Ellis, E.C. and Ramankutty, N. (2008) Putting people in the map: Anthropogenic biomes of the world. *Frontiers in Ecology and the Environment* 6, 439–447. DOI: 10.1890/070062.

Ferraguti, M., Martínez-de La Puente, J., Roiz, D., Ruiz, S., Soriguer, R. *et al.* (2016) Effects of landscape anthropization on mosquito community composition and abundance. *Scientific Reports* 6, 29002. DOI: 10.1038/srep29002.

Gascon, M., Triguero-Mas, M., Martínez, D., Dadvand, P., Forns, J. *et al.* (2015) Mental health benefits of long-term exposure to residential green and blue spaces: A systematic review. *International Journal of Environmental Research and Public Health* 12, 4354–4379. DOI: 10.3390/ijerph120404354.

Jongman, R.H.G., Külvik, M. and Kristiansen, I. (2004) European ecological networks and greenways. *Landscape and Urban Planning* 68, 305–319. DOI: 10.1016/S0169-2046(03)00163-4.

Li, Z., Xu, S. and Yao, L. (2018) A systematic literature mining of Sponge City: Trends, foci and challenges standing ahead. *Sustainability* 10, 1182–1194. DOI: 10.3390/su10041182.

McAlexander, T.P., Gershon, R. and Neitzel, R.L. (2015) Street-level noise in an urban setting: Assessment and contribution to personal exposure. *Environmental Health* 14, 18. DOI: 10.1186/s12940-015-0006-y.

Molina, L.T. (2021) Introductory lecture: Air quality in megacities. *Faraday Discussions* 226, 9–52. DOI: 10.1039/D0FD00123F.

Romanenko, V.N. (2011) Long-term dynamics of population density and species composition of pasture ixodid ticks (Parasitiformes, Ixodidae) in anthropogenic and natural areas. *Entomological Review* 91, 1190–1195. DOI: 10.1134/S0013873811090132.

Yang, W., He, J., He, C. and Cai, M. (2020) Evaluation of urban traffic noise pollution based on noise maps. *Transportation Research Part D. Transport and Environment* 87, 102516. DOI: 10.1016/j.trd.2020.102516.

Chapters 4, 5

Bassi, M. and Chiatante, D. (1976) The role of pigeon excrement in stone biodeterioration. *International Biodeterioration Bulletin* 12(3), 73–79.

Haag-Wackernagel, D. (2005) Parasites from feral pigeons as a health hazard for humans. *Annals of Applied Biologists* 147, 203–210. DOI: 10.1111/j.1744-7348.2005.00029.x.

Iorio, O.D., Turienzo, P., Masello, J. and Carpintero, D.L. (2010) Insects found in birds' nests from Argentina. *Cyanoliseus patagonus* (Vieillot, 1818) [Aves: Psittacidae], with the description of *Cyanolicimex patagonicus*, gen. n., sp. n., and a key to the genera of Haematosiphoninae (Hemiptera: Cimicidae). *Zootaxa* 2728, 1–22. DOI: 10.5281/zenodo.200077.

Lato, K.A., Madigan, D.J. and Veit, R.R. (2021) Closely related gull species show contrasting foraging strategies in an urban environment. *Scientific Reports* 11, 23619. DOI: 10.1038/s41598-021-02821-y.

Li, H. and Wilkins, K.T. (2014) Patch or mosaic: Bat activity responds to fine-scale urban heterogeneity in a medium-sized city in the United States. *Urban Ecosystems* 17, 1013–1031. DOI: 10.1007/s11252-014-0369-9.

Peacock, K.A. (2011) The three faces of ecological fitness. *Studies in History and Philosophy of Biological and Biomedical Sciences* 42, 99–105. DOI: 10.1016/j.shpsc.2010.11.011.

Robertson, B.A. and Hutto, R.L. (2006) A framework for understanding ecological traps and an evaluation of existing evidence. *Ecology* 87(5), 1075–1085. DOI: 10.1890/0012-9658(2006)87[1075:AFFUET]2.0.CO;2.

Seress, G. and Liker, A. (2015) Habitat urbanization and its effects on birds. *Acta Zoologica Academiae Scientiarum Hungaricae* 61, 373–408. DOI: 10.17109/AZH.61.4.373.2015.

Shirai, M. and Sasano, K. (2020) Bird-dropping nuisance under power lines used as perching and roosting sites by crows *Corvus*. *Reports of the City Planning Institute of Japan* Available at, 294–296. Available at: https://cpij.or.jp/com/ac/reports/19_294.pdf (accessed 10 January 2022).

Tamada, K. and Fukamatsu, N. (1992) Seasonal changes in the number and age composition of crows captured by multi-traps. *Japanese Journal of Ornithology* 40, 79–82.

Tang, Q., Vargo, E.L., Ahmad, I., Jiang, H., Varadínová, Z.K. *et al.* (2024) Solving the 250-year-old mystery of the origin and spread of the German cockroach, *Blattella germanica*. *Proceedings of the National Acadamy of Sciences of the United States of America* 121, e2401185121. DOI: 10.1073/pnas.2401185121.

Warren, R.J. (1997) The challenge of over-abundance in the 21st Century. *Wildlife Society Bulletin* 25, 213–214.

Chapters 6, 7, 8

Aak, A., Hage, Ø.M., Byrkjeland, R., Lindstedt, H.H., Ottesen, P. *et al.* (2021) Introduction, dispersal, establishment and societal impact of the long-tailed silverfish *Ctenolepisma longicaudatum* (Escherich,1905) in Norway. *BioInvasions Records* 10(2), 483–498. DOI: 10.3391/bir.2021.10.2.26.

Bertone, M.A., Leong, M., Bayless, K.M., Malow, T.L.F., Dunn, R.R. *et al.* (2016) Arthropods of the great indoors: Characterizing diversity inside urban and suburban homes. *PeerJ* 4, e1582. DOI: 10.7717/peerj.1582.

Evans, T.A., Forschler, B.T. and Grace, J.K. (2013) Biology of invasive termites: A worldwide review. *Annual Review of Entomology* 58, 455–474. DOI: 10.1146/annurev-ento-120811-153554.

Feng, A.Y.T. and Himsworth, C.G. (2014) The secret life of the city rat: A review of the ecology of urban Norway and black rats (*Rattus norvegicus* and *Rattus rattus*). *Urban Ecosystems* 17. DOI: 10.1007/s11252-013-0305-4.

Janowiecki, M.A., Austin, J.W., Szalanski, A.L. and Vargo, E.L. (2021) Identification of Reticulitermes subterranean termites (Blattodea: Rhinotermitidae) in the Eastern United States using inter-simple sequence repeats. *Journal of Economic Entomology* 114(3), 1242–1248. DOI: 10.1093/jee/toab028.

Leong, M., Bertone, M.A., Savage, A.M., Bayless, K.M., Dunn, R.R. *et al.* (2017) The habitats humans provide: Factors affecting the diversity and composition of arthropods in houses. *Scientific Reports* 7, 15347. DOI: 10.1038/s41598-017-15584-2.

Pascual, J., Franco, S., Bueno-Marí, R., Peracho, V. and Montalvo, T. (2019) Demography and ecology of Norway rats, *Rattus norvegicus*, in the sewer system of Barcelona (Catalonia, Spain). *Journal of Pest Science* 93, 711–722. DOI: 10.1007/s10340-019-01178-6.

Puckett, E.P., Orton, D. and Munshi-South, J. (2020) Commensal rats and humans: Integrating rodent phylogeography and zooarchaeology to highlight connections between human societies. *BioEssays* 42, e1900160. DOI: 10.1002/bies.201900160.

Shan, Y., Wu, W., Fan, W., Haahtela, T. and Zhang, G. (2019) House dust microbiome and human health risks. *International Microbiology* 22(3), 297–304. DOI: 10.1007/s10123-019-00057-5.

Chapter 9

Alkishe, A., Raghavan, R.K. and Peterson, A.T. (2021) Likely geographic distributional shifts among medically important tick species and tick-associated diseases under climate change in North America: A review. *Insects* 12(3), 225. DOI: 10.3390/insects12030225.

Buczkowski, G. and Bertelsmeier, C. (2017) Invasive termites in a changing climate: A global perspective. *Ecology and Evolution* 7, 974–985. DOI: 10.1002/ece3.2674.

Dhang, P. (2016) Climate change and urban pest management. In: Dhang, P. (ed.) *Climate Change Impacts on Urban Pests*. CAB International, Wallingford, UK, pp. 43–64.

Festa, F., Ancillotto, L., Santini, L., Pacifici, M., Rocha, R. *et al*. (2023) Bat responses to climate change: A systematic review. *Biological Reviews* 98, 19–33. DOI: 10.1111/brv.12893.

Fick, S.E. and Hijmans, R.J. (2017) WorldClim 2: New 1-km spatial resolution climate surfaces for global land areas. *International Journal of Climatology* 37(12), 4302–4315. DOI: 10.1002/joc.5086.

Gilbert, L. (2021) The impacts of climate change on ticks and tick-borne disease risk. *Annual Review of Entomology* 66, 373–388. DOI: 10.1146/annurev-ento-052720-094533.

Hart, C., Schad, L.A., Bhaskar, J.R., Reynolds, E.S., Morley, C.P. *et al*. (2022) Human attachment site preferences of ticks parasitizing in New York. *Scientific Reports* 12, 20897. DOI: 10.1038/s41598-022-25486-7.

Malone, M., Ivanov, K., Taylor, S.V. and Roger Schürch, R. (2023) Fast range expansion of the red imported fire ant in Virginia and prediction of future spread in the US. *Ecosphere* 14(8), e4652. DOI: 10.1002/ecs2.4652.

Parr, C.L. and Bishop, T.R. (2022) The response of ants to climate change. *Global Change Biology* 28, 3188–3205. DOI: 10.1111/gcb.16140.

Richardson, J.L., McCoy, E.P., Parlavecchio, N., Szykowny, R., Beech-Brown, E. *et al*. (2025) Increasing rat numbers in cities are linked to climate warming, urbanization, and human. *Science Advances* 11(5), eads6782. DOI: 10.1126/sciadv.ads6782.

Subject Index

Species Index

The use of scientific names in the book is limited. The common names of insects, other arthropods, and vertebrate species are officially linked to a scientific name, but the use of the scientific name does not add more information.

American cockroach, *Periplaneta americana* 46
American dog tick, *Dermacentor variabilis* 112
Anobiid powderpost beetle, *Euvrilletta peltate* 104
Argentine ant, *Linepithema humile* 67

Bed bug, *Cimex lectularis* 41
Bird feeding bug, *Psitticimix uritui* 33
Black widow spider, *Latrodectus mactans* 70
Black-legged deer, tick, *Ixodes scapularis* 114
Brown dog tick, *Rhipicephalus sanguineus* 114
Brown rat (Norway rat), *Rattus norvegicus* 75
Brown recluse spider, *Loxosceles recluse* 69
Brownbanded cockroach, *Supella longipalpa* 49

Carpenter ants, *Camponotus* sp. 93
Casemaking clothes moth, *Tinea pellionella* 101
Cat flea, *Ctenocephalides felis felis* 39
Cigarette beetle, *Lasioderma serricorne* 88
Clothes moths, *Tinea* spp. 90
Common paper wasp, *Polistes exclama*
Common European yellowjacket, *Vespula vulgaris* 55
Common N. A. yellowjacket, *Vespula alascenis* 55
Dark eye fruit fly, *Drosophila funebris* 74
Deer mice, *Peromyscus* spp. 113
Deer tick, *Ixodes scapularis* 114
Domestic house spider, *Tegenaria domestica* 68
Drugstore beetle, *Stegobium paniceum* 88
Drywood termite, *Cryptotermes brevis* 64
Dust mite, *Dermatophagoides farinae*, *D. pteronyssinus* 79

Eastern subterranean termite, *Reticulitermes flavipes* 63
European deer tick, *Ixodes ricinus* 114
European paper wasp, *Polistes dominula* 55
European starling, *Sturnus vulgaris* 28

Formosan termite, *Coptotermes formosanus* 63
Fruit flies, *Drosophila* spp. 73

German cockroach, *Blattella germanica* 44
German yellowjacket, *Vespula germanica* 54
Giant northern termite, *Mastotermes darwiniensis* 64
Granary weevil, *Sitophilus granaries* 87
Gray squirrel, *Sciurus carolinensis* 33
Green bottle fly, *Lucilia sericata* 70
Gulls, *Larus* spp. 30

House fly, *Musca domestica* 71
House mosquito, *Culex pipiens* 83
House mouse, *Mus musculus* 78

Indian meal moth, *Plodia interpunctella* 90

Little brown bat, *Myotis lucifugus* 34
Lone star tick, *Amblyomma americanum* 113
Long-tailed silverfish, *Ctenolepisma longicaudata* 103
Lyctid powderpost beetles, *Lyctus brunneus*, *L. planicollis* 103

.

www.ingramcontent.com/pod-product-compliance
Lightning Source LLC
Chambersburg PA
CBHW040136200326

41458CB00025B/6284